THE DEFINITIVE GUIDE

Astronomy

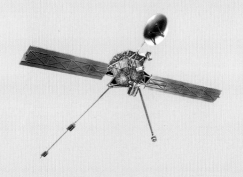

THE DEFINITIVE GUIDE

Astronomy

Robert Burnham, Alan Dyer, Jeff Kanipe

Consultant Editor
Robert Burnham

Sky Maps
Wil Tirion

FOG CITY PRESS

Published in 2003 by Fog City Press
814 Montgomery Street
San Francisco, CA 94133 USA

© 2002 Weldon Owen Pty Limited

FOG CITY PRESS
Chief Executive Officer: John Owen
President: Terry Newell
Publisher: Lynn Humphries
Managing Editor: Janine Flew
Coordinating Designer: Helen Perks
Editorial Coordinators: Sarah Anderson, Tracey Gibson
Editorial Assistant: Kiren Thandi
Production Managers: Helen Creeke, Caroline Webber
Production Coordinator: James Blackman
Sales Manager: Emily Jahn
Vice President International Sales: Stuart Laurence
European Sales Director: Vanessa Mori

WELDON OWEN PUBLISHING
Publisher: Sheena Coupe
Creative Director: Sue Burk
Project Editor: John Mapps
Designer: Lucy Bal
Picture Research: Annette Crueger

Color reproduction by Colourscan Co Pte Ltd
Printed by Tien Wah Press Pte Ltd
Printed in Singapore

ISBN 1 876778 87 3

Library of Congress Cataloging-in-Publication Data is available

A Weldon Owen Production

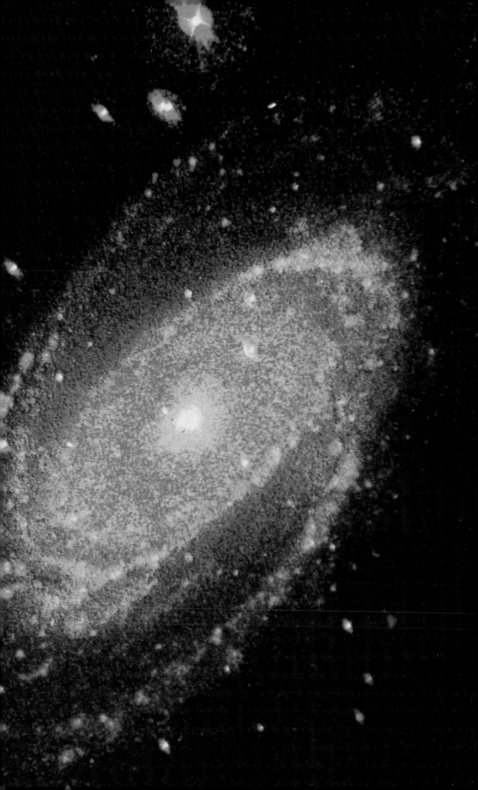

Becoming a
Stargazer

We can all delight in the wonders of the night sky just by gazing upward. A pair of binoculars opens up new vistas, and a telescope takes you into deep space. This chapter gives a rundown on these varied ways of observing the heavens and includes practical tips on finding the best binoculars and telescopes.

"Light from distant places has made the journey to earth, and it falls on these new eyes of ours, the telescopes..."

Michael Rowan-Robinson
British cosmologist, b.1942

Naked-eye Astronomy

Planetary conjunctions, eclipses, auroras, artificial satellites—these and other wondrous sights can be enjoyed without spending anything on equipment.

A prime pursuit of naked-eye astronomy is also the single most important step to exploring the sky: learning to identify the brighter stars and constellations. Spend a few clear nights through the year using the monthly star charts beginning on p. 260 to identify the brightest stars. Use these as guideposts to pick out the main constellation patterns of each month. Find a few key constellations and the rest of the sky begins to fall into place like a giant jigsaw puzzle.

Celestial Fireworks
The shimmering curtain of the aurora borealis, backed by a Full Moon, lights up the northern skies. The Southern Hemisphere equivalent is the aurora australis. Auroras occur when charged particles from active areas on the Sun interact with Earth's atmosphere.

AURORAS AND OTHER LIGHTS IN THE SKY

Observers at high northern or southern latitudes are treated to occasional displays of auroras, the northern or southern lights. These waving curtains of colored light are best seen with the unaided eye. For more information on auroras, see p. 128.

A brief streak of light across the night sky is probably a meteor, or "shooting star." At certain predictable times of the year, Earth sweeps through the

ail of dust from a comet, and meteor shower results. For more information on meteors, see p. 146.

Meteors last just seconds, but you see a star moving more slowly across the sky chances are it's a satellite. Hundreds of satellites now orbit the planet. They show up at twilight as they catch the sunlight in their orbits

hundreds of miles above Earth. The International Space Station, for example, can appear brighter than any natural star in the heavens as it slowly glides across the sky from west to east.

CONJUNCTIONS AND ECLIPSES
The planets are the sky's wandering "stars." Every few months, the motion of the planets brings two or more close together for a mutual gathering, or conjunction. Sometimes the Moon also shines nearby. Brilliant Venus sparkling near the crescent Moon is one of the sky's best naked-eye sights.

The motion of the Moon occasionally takes it across the face of the Sun, for a solar eclipse, or through our planet's shadow, for an eclipse of the Moon. Both events are highlights of naked-eye astronomy. For more information on eclipses, see p. 80.

Three-in-one
Above left: Three planets get together near the Moon to make absorbing naked-eye viewing. This conjunction shows bright Venus at center right, with Mars above it and Jupiter to Mars' left.

Bright Stars
Above right: Whatever hemisphere you live in, Orion's pattern of bright stars (at lower left) makes it one of the most readily recognizable constellations (it is pictured here as southern observers see it). Sirius, brightest star in the sky, shines at top left.

DEVELOPING NIGHT VISION
You will see more if your eyes are dark adapted and not blinded by the glare of house and street lights. It may take up to 30 minutes for your eyes to dilate and for chemical changes to take place in your retina that allow you to see in the dark. Preserve your night vision by using only dim red lights to read star charts and notebooks. Flashlights sold for astronomy that use red LEDs are best, but any pocket flashlight with the bulb covered in red cellophane will work.

Seeing Stars Under a Dark Sky

Unshielded street lighting paints a glaring skyglow above every city and town. Fortunately, the Moon and planets remain unaffected by this light pollution to provide wonderful targets for city-bound stargazers. The sharpest planet views come on nights of good "seeing," when the air is free of image-blurring turbulence and heat waves.

For the best views of the Milky Way, its star clusters and nebulas, and of faint galaxies beyond, plan to travel away from city lights, perhaps to a park or conservation area. Haze-free and moonless nights (around New Moon) are best—moonlight can wash out faint objects as effectively as can streetlight.

Light Pollution
Modern civilization, with its glow of artificial lighting, puts many barriers in the way of appreciating the sky. The scale of the problem can be clearly seen in this satellite image of the night lights of Eurasia and North Africa. In the most densely populated regions, few locations can be said to be truly dark. Red patches are flares from oil production areas; purple ones indicate burning vegetation.

Starry Night
Although you can enjoy skywatching from just about any location, a dark site far from city lights is ideal. There the Milky Way's awesome array of stars can be appreciated in all its glory.

City Lights
In many big cities, such as New York, smog and light pollution combine to rob the night sky of much of its beauty. The Moon shines through, however.

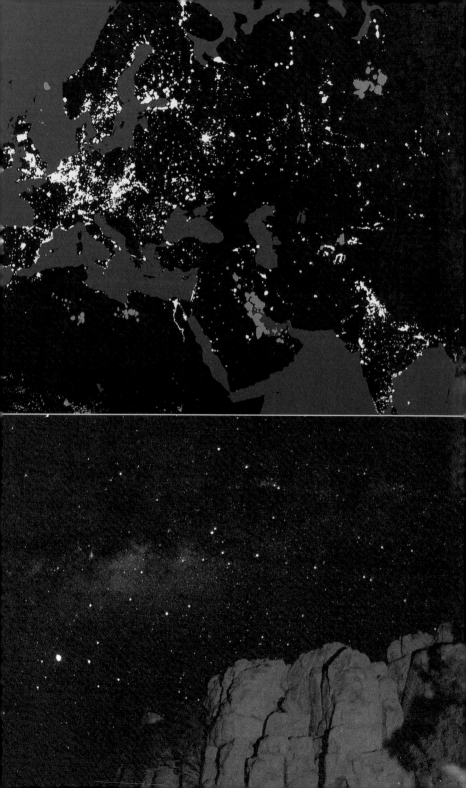

Choosing and Using Binoculars

Would you like to see sky wonders up close? Cheaper and easier to use than a high-powered telescope, binoculars provide an economy ticket to the stars.

The Right Type
Binoculars come in all sizes and magnifications. The most popular models for astronomy have front lenses 50 mm across (above). Pairs with smaller apertures, such as 35 mm, are popular for daytime pursuits such as birding.

Buying Wisely
Try out a number of binoculars before making a final decision. At the store, check the binoculars for smoothness of focus, clear views right to the edge of the field, and overall fit to your hands and eyes.

Telescopes are the icons of stargazing. Yet, surprisingly, most experienced backyard astronomers will tell you not to buy a telescope as your first piece of stargazing equipment. While telescopes provide lots of magnification, that same high power can make them difficult to aim accurately. The low power, wide field of view, and upright images of binoculars make it a snap to find celestial targets. Spend a year exploring the sky with binoculars and you will be better prepared to find objects with a telescope later.

WHAT YOU CAN SEE

The fact that you can see so much with binoculars surprises most novice stargazers. From a dark-sky site, bright star clusters, many glowing nebulas, several galaxies (including the beautiful Andromeda Galaxy), and many star-packed regions along the Milky Way are all prey for the binocular-equipped star hunter.

Binoculars can also reveal the moons of Jupiter and the largest Moon craters. A bright comet is often best seen with binoculars, as are solar and lunar eclipses, and close gatherings of the Moon and planets in the twilight.

RECOMMENDED MODELS

Binoculars are identified by a figure such as 7 x 50. The first number refers to the magnification—in this case, 7 power. Some models offer as much as 16 or 20 power, but high magnification has its drawbacks—the narrow field of view makes it hard to find objects, while shake from unsteady hands blurs images.

The second number gives the aperture, in millimeters, of each of the twin front lenses. Compared to smaller 35 mm or 42 mm models, binoculars with 50 mm lenses gather more light and provide brighter images, important for tracking down faint objects in the night sky. While binoculars

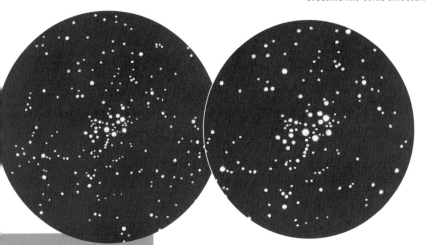

REALLY BIG BINOCULARS
Big binoculars carry specifica-
tions such as 11 x 70, 20 x 80
—even 25 x 100. These make a
poor first choice for beginners,
because they are expensive—
costing as much as a small
telescope—and very difficult to
hand-hold. But they give eye-
filling views of the sky and
should be considered by those
serious about astronomy.

with larger 60 to 80 mm lenses
re available, they are heavy and
an be a strain to hold.

The best choice is either a
7 x 50 or a 10 x 50 pair. Each
offers a good balance of power,
image brightness, and light
weight. Avoid fixed-focus or
zoom binoculars, which provide
inadequate image quality for
viewing sharp pinpoint stars.

RELIEF FOR THE EYES
An important binocular specification for many buyers is the
amount of eye relief—the distance that your eyes must be from

the eyepieces to see the whole
field of view. If you wear eyeglasses for long-distance viewing,
you may find it more convenient
to leave them on when using
binoculars. In that case, choose
binoculars with at least 18 to
20 mm of eye relief and rubber
eyecups that roll down. On such
models, even with your glasses
on, your eyes can remain com-
fortably far back from the eyepiece and still see the entire field.
With eyeglasses off, extending the
eyecups accommodates your eyes
at the correct distance.

FIELDS OF VIEW
The majority of 7-power binoculars provide a field of view of
7 degrees, enough to take in the
Pointer Stars of the Big Dipper's
bowl, or all of Crux, the Southern
Cross. Higher power 10x models
usually have a smaller field of
view, perhaps 5 degrees, still
sufficient to provide impressive
sky views.

With some binoculars, the field
is marked in degrees. If not, they
give the field as, for example,

Differing Views
One of the great binocular
sights is the Pleiades
cluster in the constellation
Taurus. These two views of
the Pleiades show the
relationship between field
of view and magnification.
Most 7x binoculars show
about 7 degrees of sky
(above left). Higher-power
10x models magnify the
image more, but usually
show only 5 degrees of
sky (above right).

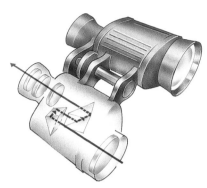

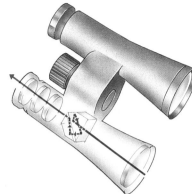

Porro and Roof
The shape of porro-prism binoculars (above left) results from the prism arrangement which produces an image that is upright and the right way round. Roof-prism types (above right) use a different prism arrangement to get the same result.

367 feet at 1,000 yards, or 125 m at 1,000 m. Divide the field by 52.4 (feet at 1,000 yards) or by 17.5 (meters at 1,000 m) to get the field in degrees. In the example above, the angular field of view is 7 degrees.

Some binoculars offer wide-angle eyepieces that provide a larger field of view than normal, perhaps 7 to 8 degrees on a 10x model. Image quality often suffers in such models, with stars at the edge of the field becoming distorted. Wide-angle binoculars often sacrifice eye relief as well. You have to pay a premium for a model that offers comfortable eye relief, wide fields, and fine image quality across the entire field.

TYPES OF PRISMS
Binoculars use one of two styles of prisms. Roof prisms provide a compact straight-through or H-shaped body. Porro prisms provide the classic zigzag or N-shaped body. For astronomy,

Binocular Testing
Factory workers test binoculars in 1939. Try before you buy is a good rule with binoculars.

LENS COATINGS: CHECKING FOR QUALITY
Another factor that sets one pair of binoculars apart from another is the type and extent of coatings applied to the lenses. Lens coatings increase light transmission and contrast. Low-cost models have at least a single layer coating on key lens surfaces, but the best binoculars feature multi-layer coatings on all optical surfaces. To check, inspect a pair by peering down the front lenses. If you see multiple bright white reflections on the internal optics, the model in question is minimally coated. In a fully multi-coated pair, the few reflections that are visible will appear dim and tinted deep green, blue, or purple.

porro prisms are usually pref-
erable; roof prisms on low-cost
models can yield annoying spikes
of light off bright stars, although
premium models work well. In
porro-prism models, units with
BAK4 glass provide a more

fully-illuminated field than do
models with cheaper BK7 glass.

HOLD STEADY!

Holding binoculars as steady as
possible improves the view, allowing you to see fainter stars and
resolve finer detail. A simple
way to ensure steadiness is to
lie on a deck chair and prop
your arms and binoculars
on the chair's arms. As an
alternative, many binoculars can
be attached to a camera tripod,
but make sure the tripod is a
sturdy one.

Tripods help with steadiness,
but are not much good for scanning directly overhead. For comfortable views of objects high in
the sky, consider a stand with
cantilever arms, which swing the
binoculars away from the tripod
allowing you to stand beneath
the binoculars.

Jewel Box

Star clusters make excellent
targets for those using
binoculars. This is the Jewel
Box, justly famous for its
attractive contrasting colors
formed into a compact,
A-shaped cluster. Views of
this object benefit from the
10x to 20x magnification of
large binoculars.

Tripod

Tripods are essential with
binoculars as big as these
11 x 70s. But even smaller
models give better views
when you brace them or
use a tripod.

Choosing and Using a Telescope

Buying a telescope involves striking a satisfactory balance
between such factors as power, image brightness, and portability
on the one hand, and cost on the other.

Compared to binoculars, the much higher magnification of a telescope resolves craters on the Moon as small as a half-mile across (about 1 km), clouds swirling in the atmosphere of Jupiter, and not just the rings of Saturn but divisions in the rings. A telescope can also range well beyond the Solar System to reveal star clusters, nebulas, and galaxies that are too small and faint to show up in binoculars.

Telescope Savvy
This reflector is marked with a checklist of quality features that applies to all telescope types. Look for a telescope with all metal and wood construction, with minimal use of plastic.

Interchangeable eyepieces, with 25 mm (low power) and 9 or 10 mm (high power) eyepieces supplied

A good finderscope, at least a 6 x 30 model (6 power, 30 mm aperture)

A smooth focuser with no wobble or backlash

Slow-motion controls (manual or electric) on both axes

A solid mount that damps out vibrations quickly

APERTURE AND MAGNIFICATION

When shopping for a telescope, the most important specification is the telescope's aperture—the diameter of its main lens or mirror. The larger the aperture, the brighter and sharper the images. A 6 inch (150 mm) telescope will provide images twice as sharp and four times brighter than will a 3 inch (75 mm) telescope.

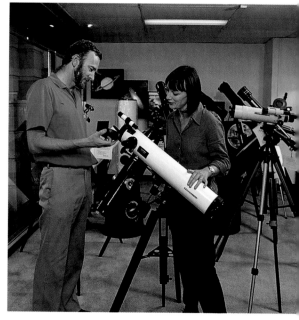

The least important specification is the telescope's magnification. By changing eyepieces, any telescope can be made to magnify any amount, although rarely will you need to employ more than 150x to 200x. Avoid any telescope sold on the basis of how much it magnifies.

TELESCOPE TYPES
—PROS AND CONS

Refractors with an aperture of more than 4 inches (100 mm) are large, heavy, and expensive, so this type of telescope is most popular in the entry-level 2.4 inch (60 mm) to 3.5 inch (90 mm) aperture sizes. These models produce sharp images of the Moon and planets, but their small apertures limit their use for hunting faint, deep-sky targets.

Reflectors, or Newtonian telescopes, lack the expensive lenses of refractors, allowing a 4.5 inch (110 mm) or 6 inch (150 mm) reflector to sell for the same price as a smaller refractor. Because of their greater aperture, reflectors are a good choice for viewing nebulas and galaxies from a dark-sky site.

Catadioptric telescope designs, such as Maksutov-Cassegrains and Schmidt-Cassegrains, offer fine optics and generous light-gathering power in compact tubes much shorter than most refractors and reflectors.

Shopping Rule
When shopping for a telescope, try to visit a range of specialist stores.

Three Types
Newtonian reflectors (near left) use a mirror to collect light, while refractors (far left) use a lens. Catadioptric types, such as the Schmidt-Cassegrain (facing page) use mirrors and a corrector lens.

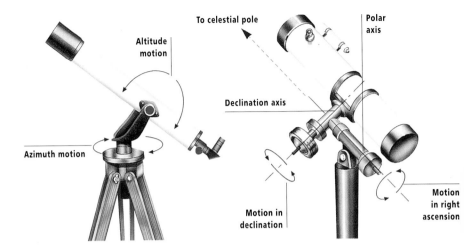

To celestial pole

Polar axis

Altitude motion

Declination axis

Azimuth motion

Motion in declination

Motion in right ascension

Know Your Mount

Your telescope's mount is as important as its optics. The altazimuth mount (above left) is widely used on small refractors and is also found on many of the amateur reflectors. The equatorial mount is used on more sophisticated telescopes because it permits motorized tracking of the stars. Equatorial mounts come in several forms, including the German-style (above right).

Dobsonian Mount

This 6 inch (150 mm) reflector uses a Dobsonian mount, which is derived from the altazimuth type. Its ease of use makes it a great choice for those new to astronomy.

TELESCOPE MOUNTS

A flimsy mount or spindly tripod will produce images that never stop bouncing, making for frustrating viewing. Mounts come in several varieties.

Altazimuth mounts provide simple up-down and side-to-side motions. The best are equipped with slow-motion controls to allow fine pointing adjustments. These mounts are easy to set up but cannot follow the stars with one simple motion. Even at low power, you will need to constantly adjust both axes to keep your target in view.

Dobsonian mounts are a variation of the altazimuth design. These simple and low-cost wooden mounts riding on Teflon pads provide motions so smooth that occasional gentle nudges are all that are needed to keep objects centered.

Equatorial mounts can track the stars by turning around one axis of rotation. These sophisticated mounts either come with

motors to do the turning, or have battery-operated motors available as options.

Fork mounts are standard on many catadioptric telescopes. When tilted over on a matching wedge or tripod, these turn like an equatorial mount to automatically follow the stars.

WHAT YOU CANNOT SEE

While bigger telescopes can show you more, be warned: no telescope will show nebulas and galaxies with the vivid colors depicted in long-exposure and computer-enhanced photographs. The human eye is simply not sensitive enough to see much color in faint deep-space objects, even through a large telescope. Do not expect to see the flags or footprints on the Moon— no telescope on Earth is powerful enough to show these.

THE BEST CHOICES FOR THE BEGINNER

A 6 inch (150 mm) or 8 inch (200 mm) reflector teamed with a Dobsonian mount is one of the best buys for the serious stargazer. You get generous aperture on a rock-steady mount that is simple to set up.

At a similar cost, 4.5 inch (110 mm) reflectors on equatorial mounts have long been popular as starter telescopes. The mounts are more complex to set up (they must be aligned to the celestial pole to track properly), but do make it easier to follow objects. Look for models with "heavy-duty" or premium-grade mounts.

Many people prefer to select a refractor because it needs less maintenance than a reflector. A 2.7 inch (70 mm) to 3.5 inch (90 mm) refractor, provided it has a solid altazimuth or equatorial mount, offers sharp optics in a rugged package that rarely needs upkeep. Avoid 2.4 inch (60 mm) refractors, as few models in this size offer good-quality mounts and fittings.

If portability is very important, a catadioptric makes a good, though more costly, choice. Popular models include 3.5 inch (90 mm) and 5 inch (127 mm) Maksutov- and Schmidt-Cassegrain telescopes. The 8 inch (200 mm) Schmidt-Cassegrain models have long been a top-selling choice of amateur astronomers looking for sizable aperture in a compact telescope.

On Target
Modern telescopes range from simple, large "light buckets" that bring in views of distant galaxies to computer-controlled types like this one, which use built-in databases to find just about anything in the night sky.

Deep-Sky Scope
With an aperture 24 inches (600 mm) in diameter, this homemade reflector is a world away from small, entry-level scopes, but it shows what an enthusiastic amateur can aspire to.

Computerized Telescope
A computer-controlled telescope makes it easy to locate even the most obscure galaxy—all at the push of a button.

COMPUTER DRIVES

An increasing number of affordable telescopes are sold with electric motors and handheld computers. After an initial aiming and alignment on two stars, these "Go To" telescopes can aim themselves to any of hundreds of objects in their databases, or under the command of a separate laptop or palmtop computer. This type of technology works most accurately and reliably on premium telescopes with high-quality motors and gears.

FINDER AIDS

The finder is the small sighting scope that sits on the side of the telescope, and a good-quality one is essential to locating objects. Many telescope owners replace poor 5 x 24 finders with a better grade of 6 x 30 finderscope or, if their telescope can handle it, a larger 8 x 50 model.

In recent years sighting aids called reflex finders have become popular. In these devices you look through a window to see a red bull's-eye projected onto the background sky, showing exactly where your telescope is aimed.

ADDING EYEPIECES

You change magnifications by changing eyepieces. A basic set of three eyepieces might include a 25 mm (for low-power views of deep-sky objects), an 18 to 12 mm (for medium-power views of the Moon and star clusters), and a 10 to 7 mm (for high-power views of planets and double stars). Invest in quality eyepieces, such as Kellners and Modified Achromats. A step up

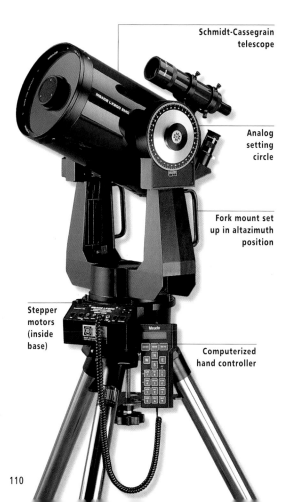

Schmidt-Cassegrain telescope

Analog setting circle

Fork mount set up in altazimuth position

Stepper motors (inside base)

Computerized hand controller

Accessories
Useful accessories include eyepieces (back), glass filters (center) to provide clearer images of some objects, a Barlow lens to increase the power of any eyepiece used with it (front), and an adaptor to attach a camera to the telescope (right).

takes you to Plössl eyepieces, perhaps the most popular style sold today.

A very useful alternative to high-power eyepieces is to add a Barlow lens. One of these will double or triple the power of any eyepiece inserted into it, while retaining the comfortable eye relief inherent in all low-power eyepieces.

CALCULATING POWER
To work out how much power a particular eyepiece provides, divide the eyepiece's focal length (usually marked in millimeters on the barrel) into the focal length of the telescope (usually marked on the tube or given in the instruction manual). For example, a 20 mm eyepiece inserted into a telescope with a focal length of 1000 mm gives 1000 divided by 20 = 50 power. A 10 mm eyepiece would give 100x, while a 5 mm eyepiece would yield 200x. The smaller the number on the eyepiece, the higher the power it provides.

Child's Play
With their sharp eyesight and insatiable curiosity, children make excellent skywatchers, and the gift of a telescope can lead to a lifelong hobby. Whatever telescope you choose, make certain it has a sturdy mounting, and test it for wobbles yourself.

HOW TO SEE THE MOST
Looking through a telescope is an acquired skill. It may take a few months of viewing before your eyes become trained to see the finest details on the small disks of planets. To see faint nebulas and galaxies, try looking off to one side to place the image on the more light-sensitive outer portion of your eye's retina. This averted vision will often reveal otherwise invisible deep-sky denizens. Accessory light pollution filters block the wavelengths from artificial lighting to enhance the subtle contrast of faint nebulas against the background sky.

How to Capture the Sky on Film

Capturing the stars is simple. Load a camera with fast ISO 400 to 800 film, place it on a tripod, and focus the lens to infinity. Set the lens aperture to f/2.8, then frame a constellation. Use a cable release to hold the shutter open for 20 to 40 seconds and you will record as many stars as your eyes can see. Recording more stars or faint nebulosity requires a dark site and bolting the camera to a motor-driven equatorial mount that can accurately track the stars during a 5- to 15-minute exposure. This piggy-back technique is simple, but the results can be spectacular.

Moonlit Scene
On Full Moon nights, use exposures of 40 seconds at f/2.8 on ISO 400 film to capture stars shining above moonlit nightscapes.

Star Trails
From a dark site, locking the shutter open for 10 to 60 minutes creates star trail portraits as Earth's motion causes stars to streak across the film.

Aiming at the Sky
Astrophotography does not require elaborate, high-tech equipment. In fact, cameras with simple mechanical shutters work more reliably than electronic models whose batteries can die during long exposures.

Astronomy as a Hobby: Where to Now?

Whether you become a serious observer or not, you will be able to share your interest in the sky with amateur astronomers in your own home town or city and around the world.

Comet Watch
Observing comets is a captivating experience for any astronomer. This is Comet Hale-Bopp in 1997.

Star Parties
Star parties are great opportunities to meet fellow skywatchers and try out a range of instruments.

Like the universe it studies, the hobby of amateur astronomy has no limits. You can simply lie back and enjoy the beauty of the night sky, or probe deepest space with telescopes. You can capture images of celestial wonders, or embark on a program that will contribute data to research astronomers. Whatever your depth of interest, you will enjoy the hobby more if you share it with others.

JOIN THE CLUB

In most cities and towns you will find a club of like-minded amateur astronomers. Chances are they are affiliated with a local science or nature center, planetarium, college, or university. Astronomy clubs host regular meetings indoors as well as weekend viewing sessions at favorite dark-sky sites. These star parties provide an opportunity to meet fellow skywatchers, see a variety of telescopes in action, and get advice about what equipment to buy.

THE VIRTUAL WORLD

Through the Internet, amateur astronomers are now linked

together into many virtual clubs. Hundreds of amateurs from around the world chat together on newsgroups, while e-mail lists serve the hobby's special interests, such as astrophotography and deep-sky observing. On-line user groups for popular telescopes provide helpful hints and advice for telescope owners. The major astronomy magazines maintain websites with extensive links to other hobby-related sites as well as listings of astronomy clubs and major star parties. For website addresses, see the Resources Guide beginning on p. 416.

GETTING SERIOUS

While most backyard stargazers are content to enjoy the sky as a casual hobby, some take a more ambitious approach by conducting scientifically useful observations. Professional astronomers depend on amateurs to follow the pulsating light output of variable stars, to scan the sky for exploding nova stars, and to monitor the planets for outbreaks of storms. Amateur astronomers also routinely discover comets and asteroids, and help trace the rise and fall of meteor showers. Websites for organizations that seek reports from amateurs are listed in the Resources Guide.

Pursuing the Hobby
Among other things, an astronomy club's get-together is a chance to see just how far you can pursue the hobby.

KEEPING A JOURNAL
Even if they do not report any observations, many amateur astronomers find it rewarding to keep a logbook of their star treks. Such sky diaries record your progress as an observer and are fun to look back on to read impressions of a memorable night under the stars. Describe what you saw and the sky conditions. Make the effort to include sketches of eyepiece views. The goal is not to create works of art; rather, drawing what you see forces your eye and brain to see more, developing your observing skills still further.

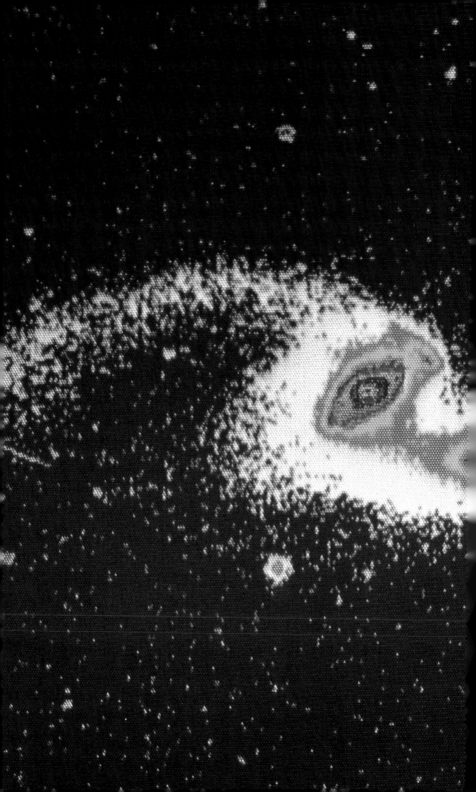

The Cosmic Journey

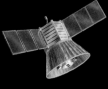

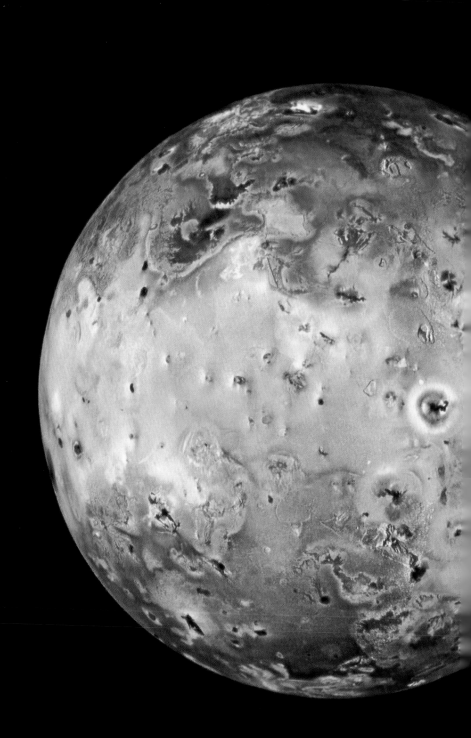

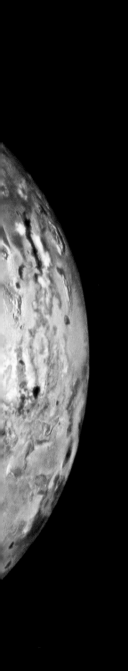

Our
Solar System

The Sun sits at the center of a diverse system that includes the nine major planets and their moons, together with millions of asteroids and innumerable other bodies. Large or small, the Solar System's members are highly individual, from Saturn and its spectacular rings to the frigid world of tiny Pluto.

"He, who through vast immensity
can pierce,
See worlds on worlds comprise one universe,
Observe how system into system runs,
What other planets circle other suns,...
May tell why Heaven has made us as
we are."

An Essay on Man, Alexander Pope
English poet, 1688–1744

The Sun's Family

The Solar System comprises our tiny, familiar corner of the universe. Five billion years ago the Sun and its disparate retinue of orbiting bodies existed only as gas and dust.

Creation Story

Astronomers are now generally agreed on how the Solar System formed. About five billion years ago, grains of material from a rotating cloud of gas and dust (1) consolidated into solid lumps of material (2). Through violent collisions with one another, these planetisimals formed larger bodies, proto-planets (3). These consolidated further through gravitational encounters and more collisions (4). Finally, after no more than 100 million years, the newborn Sun abruptly brightened, and its radiation blew away any material that had not been swept up by the planets (5).

The Solar System consists of the Sun; nine planets and their dozens of satellites; millions of asteroids, or minor planets; innumerable meteors and comets; and a plane of dust that pervades all of interplanetary space. How did all these very different objects come into being? Although many gaps in our knowledge remain, most astronomers believe that the Solar System formed less than five billion years ago from a cloud of hot hydrogen and helium gas and dust known as the solar nebula.

FROM SOLAR NEBULA TO SOLAR SYSTEM

Triggered perhaps by shockwaves from a nearby supernova, or by the gravitational disturbance of a passing star, the solar nebula began to contract and rotate. Its central region collapsed into a protostar, which began to grow. In the broad disk surrounding the protostar, particles began to condense from knots of gas. These grew into planetisimals—bodies of rock and ice a few miles across—

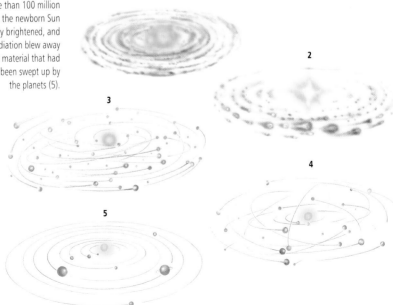

1

2

3

4

5

which collided and coalesced with one another to form larger bodies called protoplanets, precursors to the planets.

The protostar became a star when its core temperature reached 18 million°F (10 million°C) and nuclear fusion reactions began—our Sun was born. Radiation produced a powerful solar wind of charged particles, a "shock wave" that blew away remaining gas and dust grains from the Solar System.

Heating by gravitational contraction and radioactivity caused the protoplanets to melt and separate into the different internal layers we can observe in planets today. The first billion or so years was also a time of continuous bombardment by leftover debris, which battered the surfaces of the planets and their satellites. The craters of the Moon are mementos of this era.

WHEN WORLDS COLLIDE

The final stages of the Solar System's formation included catastrophic collisions between planet-size bodies. Mercury, with its unusually large core, may be the remnant of one such incident. A collision between Uranus and a planet twice as massive as Earth may have caused the severe tilt of Uranus' rotation axis. And it is now thought that the Moon formed from debris created when a body the size of Mars struck the young Earth.

Planets in Motion

The nine planets of the Solar System orbit the Sun in the same direction and almost on the same plane. WIth one exception, their elliptical paths are very nearly circular. Pluto has a highly elliptical orbit that sometimes takes it inside Neptune's orbit.

THE NATURE OF THE PLANETS

Because temperature decreases with distance from the Sun, the composition of the planets differs according to where they formed. The inner solar nebula was far

Gas Giants

The gas-giant planets are well named. Jupiter, for example, is far larger than the rocky, inner planets—so large, in fact, that more than a thousand Earths could fit inside it. And, like the other gas giants, it consists mostly of hydrogen and helium; the swirling, banded atmosphere we can see (below) is merely the outermost layer of a huge, mostly gaseous ball.

too hot for such substances as water, methane, and ammonia to exist as solids; instead, the inner planets formed from materials such as iron and silicates. The cold of the outer solar nebula allowed planets there to hold onto significant amounts of water, ice, and other elements easily destroyed by heat. The greater mass of these planets also meant that they could sweep up large quantities of hydrogen and helium, which created the voluminous atmospheres of the giant planets.

are comets, small frozen objects which normally inhabit the Solar System's outer edges. Occasionally a comet may leave these faraway regions, head toward the Sun, and show up dramatically in our skies.

Asteroids, also called the minor planets, are made up of metallic rock. Most orbit the Sun within the main belt, the region between Mars and Jupiter. Fragments of comets and asteroids are known as meteoroids, which may rain down on Earth producing the bright streaks we call meteors.

LEFTOVER BODIES

The Solar System was not completely "cleaned up" by the Sun's birth. Among the objects left over from the solar nebula

ORBITS GREAT AND SMALL

All planetary bodies, from Jupiter to the smallest particle of dust, move about the Sun in elliptical orbits. An ellipse is a kind of flattened circle with two foci positioned on opposite but equal sides of its center. In the Solar System, the Sun occupies one focus. This means that each body goes through two extreme points in its orbit, one nearest the Sun, called perihelion, and one farthest from the Sun, aphelion.

All bodies, except comets, orbit the Sun in the same direction as Earth. If you could stand well above Earth's North Pole and

Mercury: 36 million miles (58 million km)
Venus: 67 million miles (108 million km)
Earth: 93 million miles (150 million km)
Mars: 142 million miles (228 million km)
Asteroid belt

Jupiter: 484 million miles (778 million km)
Saturn: 890 million miles (1.4 billion km)
Uranus: (1.8 billion miles (2.9 billion km

OTHER SOLAR SYSTEMS

Until very recently our Solar System was *the* solar system. It is now known that other stars harbor their own retinue of planets (see p. 188). So far, only Jupiter-size bodies have been discovered, but it may not be long before the first rocky planets are detected, including, perhaps, worlds like Earth.

ok down on the Solar System, ou would see the planets moving ounterclockwise. Most of the lanets also rotate in the same irection as their orbital motion.

The planets orbit nearly in the me plane, close to that of Earth's rbital plane (the ecliptic) and the un's equator. Pluto is the odd an out, with an inclination to le plane of 17 degrees. Comets re able to approach the Sun from y direction and angle.

IVIDING THE PLANETS

lanets fall into two groups based n composition. Those nearest e Sun—Mercury, Venus, Earth, d Mars—are called terrestrial lanets because they are made up rimarily of rock and metals. piter, Saturn, Uranus, and Teptune are the jovian or gas-iant planets. They have masses least 10 times that of Earth

and are made up largely of lighter ices, liquids, and gases. Pluto is neither terrestrial nor jovian, but resembles instead the rock–ice moons of the outer planets.

Another grouping of planets is based on distance. Those with orbits inside Earth's orbit—Venus and Mercury—are called inferior, or inner, planets; those outside our orbit—Mars, Jupiter, Saturn, Uranus, Neptune, and Pluto—are the superior, or outer, planets.

Rocky Worlds
The four inner worlds of the Solar System, including Venus (above) and Mars (left), are known as the terrestrial planets because they have certain Earth-like (terrestrial) features. Chief among these is their rocky nature, solid surfaces, and relatively small size.

Solar System Distances
This diagram shows the average distances of the planets from the Sun. Compared to the outer planets, Mercury, Venus, Earth, and Mars are crowded together in the glare of the Sun.

Neptune: 2.8 billion miles (4.5 billion km) | Pluto: 3.7 billion miles (5.9 billion km)

Moons:
Companions to the Planets

Every planet, except Mercury and Venus, has at least one satellite, or moon. Some planets have miniature solar systems: Saturn has at least 30 moons, Jupiter 39 (so far), and Uranus 21. The moons orbit their host worlds in much the same way that planets go around the Sun, in elliptical orbits and with definite periods.

Moons vary greatly in size, from small mountains to planet-size bodies. Each is a world unto itself, with its own density, chemical makeup, and geological quirks. A few, such as Saturn's Titan, have an atmosphere, while others, including Jupiter's Io and Neptune's Triton, exhibit a complex geological history.

Size Compariso
A comparison of some of the Solar System
moons and planets shows how greatly th
range in size. Note how small Pluto
compared to Earth's Moon and Jupiter
Ganymede. Phobos, one of Mars' moons, is
boulder only 8.4 miles (13.5 km) acros

Different Surfac
At the surface, moons are an extreme
diverse bunch. Miranda (below), one
Uranus' satellites, is a geologic jumble, wi
fracture patterns, craters, and other feature
Europa's surface (right) is a shell of ice cris
crossed with fractures filled with fresh ic

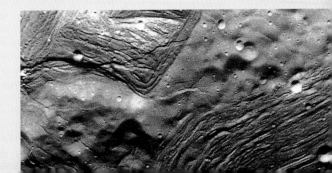

Neptune

Phobos   Pluto   Moon   Ganymede   Earth

The Sun

Our Sun is a gigantic ball of hydrogen and helium gas in which nuclear reactions create awesome amounts of energy, including the heat and light on which we on Earth depend.

Coronal Loops
The Sun is in a constant state of turmoil. Its outer atmosphere, the corona, is heated by millions of brilliant arches, called coronal loops. These fountains of hot gas may be up to 300,000 miles (480,000 km) high.

Proxima Centauri, 4.24 light-years away, is often called the closest star to Earth. In fact, our closest star is the Sun, on average 93 million miles (150 million km) away—just around the corner in cosmic terms. We may take it for granted, but this yellow dwarf is out of the ordinary in size and brightness compared to most other stars: only 2 percent of the Milky Way's stars are larger and brighter.

A DAYTIME STAR

The Sun is simply enormous. Its diameter of 865,000 miles (1,392,000 km) is nearly 110 times Earth's and 10 times that of Jupiter. It comprises 99.8 percent of the entire mass of the Solar System; the other 0.2 percent is everything else—Earth, the planets and moons, asteroids, comets, and dust.

Though it seems relatively tranquil, the Sun is a seething sphere of mostly hydrogen gas. Its source of energy is a nonstop succession of nuclear reactions in its core, where hydrogen atoms fuse to form helium. The Sun radiates a staggering amount of energy, not just as light and heat, but also several other different types of radiation, including gamma rays, ultraviolet rays, and X rays.

THE STRUCTURE OF THE SUN

The temperature at the Sun's core is 27 million°F (15 million°C). So vast is the Sun that by the time this heat reaches the photosphere—the visible surface—it has cooled to a little over 10,000°F (5800°C). Above the photosphere lies a thin, cool layer called the chromosphere, which is normally invisible. Above this is

THE SUN	
Distance from Earth	93 million miles (150 million km)
Diameter	865,000 miles (1,392,000 km)
Mass	333,000 Earth masses
Spectral class	G2 V
Surface temperature	10,000°F (5,800°C)
Visual magnitude	−26.7
Absolute magnitude	4.8

e outer atmosphere, the corona,
here the temperature rises to
an estimated 3.6 million°F
 million°C). Like the chromo-
here, the corona is visible only
uring solar eclipses or through
special filter.

UNSPOTS AND
THER SOLAR FEATURES

is in the photosphere that you
an observe signs of solar activity
hrough a small telescope
quipped with a proper solar
ter. The best-known features
e sunspots, dark regions which
e somewhat cooler than their
surroundings. Over a period of
bout 11 years, sunspot numbers
ncrease then slowly decline.
his sunspot cycle more or less
arallels the increase and decrease
the Sun's overall activity.

The most complex sunspots are
ubs of intense magnetic fields.
hese active regions can suddenly
rupt as flares—short-lived,
xtremely bright areas that release
rge amounts of charged particles
nd radiation. Flares are more
revalent during peaks in solar
ctivity and, unfortunately, none

but the very brightest are visible
without a special narrow-band
hydrogen-alpha filter.

Also visible on the photosphere
are bright cloudlike filaments
called faculae. Unlike sunspots,
these features are hotter than
their surroundings.

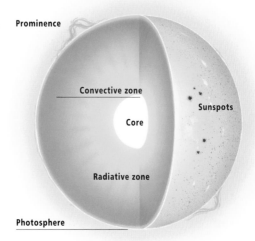

Prominence
Convective zone
Core
Sunspots
Radiative zone
Photosphere

Inside the Sun
The Sun's photosphere—
the visible surface—with
its dark spots and looping
prominences, overlies
boiling convection layers.
Deeper down, energy
streams out from the
nuclear powerhouse
at the core.

Sunspots
Sunspots arise where the
Sun's magnetic field is
concentrated, impeding the
flow of energy. A typical
sunspot (left) consists of a
dark region (the umbra)
surrounded by a lighter
region (the penumbra).
The umbra is about 3600°F
(2000°C) cooler than the
photosphere and only looks
dark in relation to its
surroundings. Spots tend
to occur in groups (below
left), which are carried
across the solar disk by
the Sun's rotation.

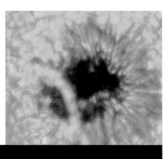

Atmospheric Fireworks Courtesy of the Sun

Space is said to be a vacuum, but that does not mean it is devoid of its own kind of "weather." The primary space-weather generator is the solar wind, made up of supersonic streams of charged particles (protons and electrons) flowing outward from the Sun. On Earth, the solar wind often manifests itself as auroras— sublime displays of shifting colored lights seen mainly in and around the polar regions. But Earth is not the only planet that experiences auroras.

The Voyager spacecraft discovered lightning displays and auroras in the upper atmospheres of both Jupiter and Saturn. As on Earth, these extraterrestrial auroras occur near the planets' poles but are much brighter.

Auroras in Spac
The Sun sends a steady stream of charge
particles, the solar wind, out into spac
causing auroras on Earth (seen from spac
top), Jupiter (below), and Saturn. The
charged particles are drawn by the plane
magnetic field to the poles where th
interact with gas molecules in the upp
atmosphere, causing them to emit ligh

Curtains of Ligh
The aurora borealis transforms the sky ov
Manitoba, Canada. Auroras can last a fe
minutes or an entire nigh

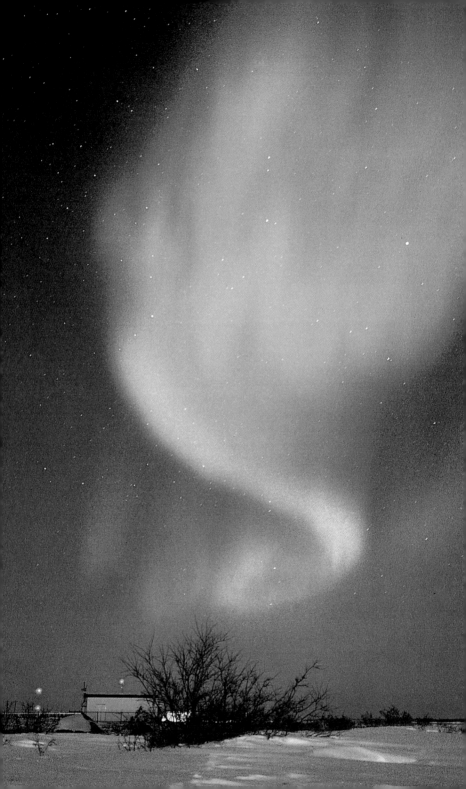

Mercury

Mercury is a small, rocky, airless world with an extreme climate—roastingly hot during the day, frigidly cold at night. Its pock-marked surface is reminiscent of the Moon's.

Mariner 10
Most of what we know about Mercury comes from Mariner 10, which made three flybys in 1974–75.

Long Day, Short Year
During its year of 88 Earth days, Mercury turns on its axis one and half times. The red marker on its surface shows that it completes only half a solar day in this time. A full solar day from noon to noon lasts 176 Earth days.

Mercury is a tricky object to see, a consequence of its position as innermost planet. It never strays farther than 28 degrees from the Sun, which makes it difficult for the planet to be seen against the Sun's glare. Nor does Mercury remain in the same part of the sky for more than a month at most—in fact, the planet can bound through as many as three constellations in the space of a month.

YEARS AND DAYS

Being just 36 million miles (58 million km) from the Sun allows Mercury to live up to its fleet-footed namesake: its orbital period is 88 days, the Solar System's shortest. Contrasting

with its year is the length of Mercury's day. If you measure a "day" by how long it takes a planet to rotate, then a day on Mercury is equivalent to slightly less than 59 Earth days, or two-thirds of its year. But if you define a day by how long it takes for the Sun to return to the same point in the sky, then a day is two Mercury years, or 176 Earth days (see diagram, below).

MERCURY—PLANET OF EXTREMES

Mercury's mass is just one-eighteenth that of Earth's, and much too small to retain an atmosphere. With the Sun so close, and lacking the protection from extremes that an atmosphere would give, Mercury has a *very* uncomfortable climate. The planet's day side reaches 800°F (430°C) by noon. The heat dissipates into space at night, when temperatures drop dramatically, bottoming out just before dawn at −280°F (−170°C). This is the widest temperature range of any of the planets.

In 1992, radar beams sent from Earth revealed the presence of ice below the surface near the poles, where the temperature remains constantly below freezing. Astronomers theorize that these underground "polar caps" may be the result of impacts by ice-rich comets on Mercury.

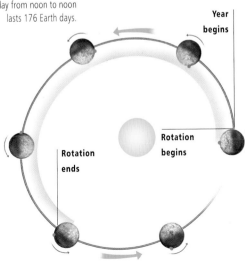

Year begins

Rotation begins

Rotation ends

Rotation begins

THE SURFACE REVEALED

In 1974, Mariner 10 saw a world that could almost double as the Moon. The entire planet is pockmarked with impact craters ranging in size from the smallest detectable, ¹/₂ mile (1 km) across, to the giant "bull's eye" of the Caloris Basin, some 830 miles (1,340 km) wide. Ridges, highlands, and lava-flooded basins complete the picture.

INSIDE MERCURY

Mercury's diameter is 3,032 miles (4,880 km), just 1.4 times larger than the Moon's, but it is 1.7 times denser, which indicates that the composition of these two bodies is very different. Scientists believe that Mercury has a unique internal structure: like a thick rind on a piece of fruit, a silicate crust only 435 miles (700 km) thick covers a huge metallic iron-nickel core making up 60 percent of the planet's total mass and filling three-quarters of its radius. The large core may have formed that way, or it may be the result of Mercury losing some of its upper layers in a massive collision.

MERCURY	
Diameter	3,029 miles (4,875 km)
Mass	0.055 Earth masses
Rotation period	58.6 days
Inclination of equator to orbit	0.5 degrees
Mean orbital speed	30 miles per second (48 km/s)
Mean distance from the Sun	36 million miles (58 million km)

Cratered Surface
Like the Moon, the surface of Mercury is dominated by impact craters and large basins; some craters also have prominent ray systems. But even its most heavily cratered regions show fewer impacts than the Moon's southern highlands. This leads scientists to assign Mercury's surface a younger age, since older surfaces typically display more craters.

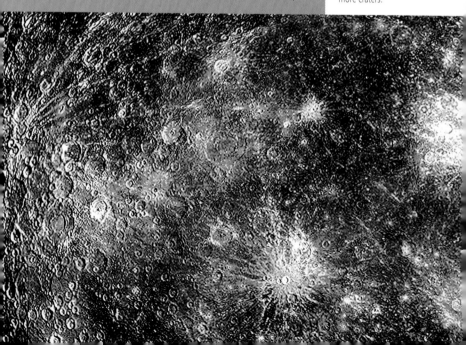

Venus

With its thick shroud of clouds, Venus kept its secrets well hidden until recent times. Space probes have found a forbidding yet fascinating world of scorching temperatures, rocky plains, and huge volcanoes.

Given its greater proximity to the Sun you would expect Venus to be somewhat warmer than Earth. In fact, it is an inferno, with a surface temperature of 890°F (470°C)—hot enough to melt lead and zinc.

Magellan
Between 1990 and 1994, the Magellan spacecraft surveyed all of Venus from orbit using radar to map the planet's cloud-covered surface in detail.

Greenhouse Effect
Venus suffers from a runaway greenhouse effect. Strong sunlight filters through the clouds and heats the surface, but the clouds and carbon dioxide in the atmosphere keep the heat from escaping back into space. Venus cannot cool down.

AN OUT-OF-CONTROL GREENHOUSE

The explanation for this disparity lies in Venus' atmosphere of carbon dioxide, a layer so dense that the pressure at the surface of the planet is nearly a hundred times that of Earth's. This thick blanket traps the Sun's heat at the planet's surface, producing an extreme greenhouse effect. Scorching temperatures result.

Cloud layers stretch upward from an altitude of 30 miles (48 km) above the surface. These are no ordinary clouds—they consist almost entirely of sulfuric acid droplets, and spread relatively uniformly around the planet.

ON THE SURFACE

Earth's crust is recycled and shaped by plate tectonics, in which the continents "ride" on huge plates. Between 1990 and 1992, radar surveys by the Magellan orbiter showed a much different process at work on Venus. Dome-like structures are evidence of upwellings of lava in the planet's mantle; these, together with widespread volcanic activity, are largely responsible for recycling the crust and sculpting a desolate landscape.

Nearly 85 percent of Venus' surface consists of flat lava plains which resemble the basaltic maria, or "seas," of the Moon. There are huge volcanoes and mountain ranges—the Maxwell Mountains rise nearly 7.5 miles (12 km) above the surrounding plains. The planet's two great landmasses are Aphrodite Terra, which girths a portion of the planet from the equator into the southern hemisphere, and Ishtar Terra, in the high latitudes of the northern hemisphere.

Venus has an estimated 900 impact craters, all bigger than about 2 miles (3 km) in diameter. The dense atmosphere has protected the surface from small

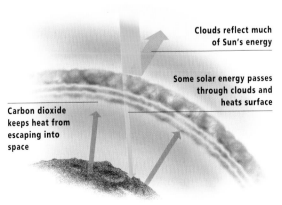

Clouds reflect much of Sun's energy

Some solar energy passes through clouds and heats surface

Carbon dioxide keeps heat from escaping into space

asteroids and comets; only large objects survived to find their way to the ground. The craters are fairly young in geological terms, at less than 500 million years old. Volcanic activity, it seems, has destroyed signs of earlier impacts.

A SLOW AND BACKWARD PLANET

Venus rotates *very* slowly—once every 243 Earth days, 18 days longer than it takes to circle the Sun. And it spins not from west to east, like the other planets, but from east to west.

This slow retrograde motion has a strange effect on the Venusian calendar. If you were on Venus, you would see the Sun rise in the west, cross the sky, and set in the east some 59 Earth days later. Perhaps an early collision between Venus and an asteroid or comet set the planet on its backward course.

Cloud Forecast
On Venus, the cloud never clears. The planet's upper atmosphere rotates every four or five days, much faster than the sluggish winds at the surface.

Folded Plateau
Ovda Regio, a plateau in one of Venus' highland areas, has undergone an extraordinary amount of folding and faulting, and looks like nothing on Earth.

VENUS	
Diameter	7,521 miles (12,104 km)
Mass	0.82 Earth masses
Rotation period	243.0 days, retrograde
Inclination of equator to orbit	177.4 degrees
Mean orbital speed	22 miles per second (35 km/s)
Mean distance from the Sun	67 million miles (108 million km)

Earth

Earth is just another planet in the Solar System. True, perhaps, but the more we learn about the other planets, the more we appreciate how special the third rock from the Sun really is.

Earth is the largest terrestrial planet and one of the most geologically active worlds in the Solar System. It is the only planet where temperatures allow surface water to exist in solid, liquid, and gaseous forms. Uniquely, too, it supports a wonderful diversity of life.

THE HOME PLANET

Several factors combine to make Earth the only planet in the Solar System hospitable to human life. It has vast oceans of liquid water and a protective atmosphere rich in oxygen. It orbits in a stable, nearly circular path, so that it is never too far from the Sun's warmth nor too close. It has

one moon—large by Solar System standards—that acts as a kind of gyroscopic stabilizer, preventing the tilt of Earth's spin axis from shifting wildly. Most important of all, it lies within a habitable zone in the Solar System: not too near or too far from the Sun—just right.

OCEANS OF WATER

On no other planet but Earth does water exist as a liquid. And we have lots of it. Saltwater oceans cover 71 percent of the planet's surface. The Pacific reaches a depth of 35,800 feet (10,900 m), but the global average is 12,500 feet (3,800 m).

As well as supporting marine life-forms of great variety and abundance, the ocean plays a vital role in sustaining ground-based life. Heat from the Sun evaporates seawater; freshwater moisture

Alien View
Swirling white clouds of water vapor (above) make Earth a brilliant beacon in the Solar System. Below the clouds, the deep blue of the oceans dominates the view. Aliens aboard an approaching spacecraft could be forgiven for making these enormous tracts of water the main focus of their studies. From up there, the continents, for all their vastness, resemble mere islands. Zooming down for a closer look, our aliens might be surprised at the ocean's ability to steadily wear away the land (right).

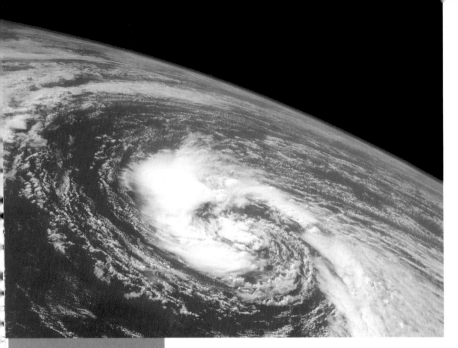

MISSING ROCKS

Although Earth formed 4.6 billion years ago, the oldest known rocks are about 3.9 billion years old. The rocks in the 700-million-year gap in the records were probably destroyed by erosion or by being pushed back into Earth's molten interior.

precipitates from this evaporation, cooling the ocean and bringing rain to the land, where it refreshes lakes, rivers, and streams, and eventually flows back to the sea. All the water in the oceans goes through this cycle once in two million years.

A BREATHABLE BLANKET THAT KEEPS US WARM

Each of us is surrounded by an envelope of air that stretches upward for over 75 miles (120 km). Earth's atmosphere is made up almost entirely of just two gases—nitrogen (78 percent) and life-sustaining oxygen (21 percent); traces of carbon dioxide are enough to create a moderate greenhouse effect, a key factor in retaining the warmth of the Sun and preventing extreme temperature swings from day to night.

The atmosphere gets thinner and colder the higher you go. The densest and warmest part, known as the troposphere, lies between the surface and an altitude of 10 miles (16 km). In this zone, air warmed by the Sun rises from the surface to be continually replaced with descending parcels of cool air. This circulatory pattern generates most of Earth's weather.

Above the troposphere lies the stratosphere. This region, which extends to approximately 37 miles (60 km) above the surface, is cold and cloudless. If you go higher still, you would encounter the

Earth's Atmosphere
Earth's atmosphere creates dramatic weather, such as this severe storm over the Pacific Ocean viewed from space. It is the interaction between the atmosphere and the Sun's heat energy that creates our weather.

The Moon

Except for a few days each month, the Moon is always visible in the sky somewhere. It has been Earth's companion for over four billion years and, according to one theory, may once have been part of our planet.

Lunar Field Trips
Apollo astronauts collected more than 1,000 pounds (450 kg) of Moon rocks.

Giant Impact
The favored explanation for the Moon's formation is the impact theory. The young Earth was struck by an object about the size of Mars. Debris from the collision circled Earth, then clumped together to form the Moon.

Despite its relatively small size of 2,160 miles (3,476 km) across, the proximity of the Moon to Earth—on average it is 238,900 miles (384,400 km) away—means that our neighbor in space is the most intensely studied celestial body. It has been the target of telescopes since Galileo, and in the space age became the only extraterrestrial world visited by humans. On six

visits by 12 Apollo astronauts, hundreds of lunar rock samples were taken, deep cores drilled into the surface, and a variety of scientific experiments performed. Unmanned space probes such as Luna, Zond, Surveyor, and Lunar Prospector have scrutinized just about every aspect of the Moon's geology. And yet a number of questions remain.

THE BIRTH OF THE MOON

Planetary scientists still debate the Moon's formation. In the past, three general theories predominated: the Moon formed in some other part of the Solar System and was gravitationally snared by Earth (the capture theory); the Moon was once part of Earth but calved from it early in their history (the fission theory); and Earth and the Moon formed together but independently, as many of the moons in the outer Solar System formed (the sister theory).

Each theory has its problems—for example, gravitational models

gue strongly against Earth being
le to capture a satellite as large
the Moon—and a fourth
eory is now favored by most
tronomers. According to the
ant impact theory, a glancing,
gh-speed collision between the
rimitive Earth and a Mars-size
ody smashed Earth's crust and
elted rock deep within its
antle. Molten rock from both
arth and the impacting body
as jettisoned into space,
ondensing into a ring of

orbiting debris. This material
coalesced, cooled, and solidified
into the Moon.

A POCKMARKED PLACE

A glance through a telescope
reveals a surface covered with
craters. Many of these formed
during the first 800 million years
of the Moon's history during a
period of intense bombardment
that pulverized and abraded the
young lunar crust. The pum-
meling culminated in the huge
impacts that created the larger
lunar basins and craters. More
recently—some hundreds of
millions of years ago—objects
smashed into the surface forming
such craters as Tycho, Copernicus,
Kepler, and Aristarchus.

The naked eye can easily
discern the Moon's mottled
surface of bright and dark

Lunar Prospector
Launched in 1994, the Lunar
Prospector probe is one of
the latest in a long line of
spacecraft to visit our near-
neighbor since 1959.

Face of the Moon
The Moon is so close to Earth
that surface features can
easily be seen by the unaided
eye. The dark patches are
maria, lava-filled basins
mistaken for dried-up seas
by early observers, and the
bright areas are highlands,
where most of the Moon's
craters are located.

Regolith
The dusty "soil," or regolith,
that covers the Moon's
surface leaves deep footprints,
as this photograph from the
Apollo 17 mission shows.
The regolith was formed by
countless tiny meteorites
striking the ground and
pulverizing rocks. Regolith
in the maria is 7 to 26 feet
(2 to 8 m) deep, while in
the highland regions it may
extend more than 50 feet
(15 m) below the surface.

occur in these areas, ranging in size from tiny pits to huge "bowls" with central mountain peaks and walled plains.

The dark regions are lightly cratered basins and lowlands called maria (singular, mare), Latin for "seas," since that is how they appeared to early sky-watchers. Prominent examples visible in the Moon's eastern hemisphere at First Quarter phase are Mare Tranquillitatis, Crisium, and Serenitatis. The maria were formed by floods of lava early in the Moon's history.

The Moon is a quiet place today, impacts being rare and volcanic activity at least a billion years in the past. Seismic activity is very low—the total amount of energy released by moonquakes is a hundred billion times less than that of earthquakes on our world. The Moon is now all but geo-logically dead.

THE MOON'S MAKEUP

The Moon is made up of largely different materials to those pre-dominating on Earth. Density is an important clue to interior composition, and planetary scientists cite the Moon's low density as evidence that it must be composed almost entirely of

Ancient Bombardment
Most of the Moon's innumerable craters formed during a cataclysmic period of bombardment that ended about 3.8 billion years ago.

patches. The bright regions consist of mountain ranges, uplands, material ejected when craters were formed, and regolith, the Moon's fine-grained soil. Dense concentrations of craters

THE MOON	
Diameter	2,160 miles (3,476 km)
Mass	0.012 Earth masses
Rotation period	27.3 days
Inclination of equator to orbit	6.7 degrees
Mean orbital speed	0.6 miles per second (1.0 km/s)
Mean distance from Earth	238,900 miles (384,400 km)

ight silicate rocks. Unlike Earth, the Moon is deficient in iron and metals and nearly devoid of materials that are gaseous at low temperatures. Modest reserves of water ice, however, were detected by the Lunar Prospector mission in the late 1990s. Prospector's data indicated that as much as 6 billion tons of ice lies buried 18 inches (40 cm) below ground in the permanently shadowed craters at the poles. This is enough water to support a city of 360,000 people for well over a century without recycling.

The Moon lacks a significant metal core and a global magnetic field, though evidence suggests that a much stronger field once existed. Researchers theorize that it dwindled because the Moon's initial core was never large or hot enough to produce the kind of powerful convection currents needed to create a strong field.

THE FAR SIDE

The Moon completes one orbit around Earth in 27.3 days, which also happens to be how long it takes to complete one rotation on its axis. The synchronicity between its orbit and rotation means that it keeps one side turned perpetually toward Earth. We may never see the far side, but spacecraft have seen it, and photographs show a similarly battered moonscape.

Valleys and Boulders
Beneath a jet-black sky, geologist–astronaut Harrison Schmitt works near a huge boulder in the Taurus-Littrow valley during the Apollo 17 mission in 1972.

Moonshine
By far the most spectacular object in the night sky, the mysterious Moon exerts a seemingly irresistible attraction to skywatchers.

Mars

Images and data sent by space probes show the Red Planet to be a forbidding desert world with a harsh climate. Even so, Mars remains a prime target for the search for life outside Earth.

Face of Mars
From space, Mars is a dry- and desolate-looking place. Across its center stretches the Valles Marineris canyon, 2,500 miles (4,000 km) long.

Polar Cap
The northern polar cap of Mars (like the southern) can be seen in backyard telescopes (see p. 246). Mostly water-ice mixed with dust, the cap sits on a "layer cake" of water-ice deposits that is several miles thick.

Mars is often cited as being the most Earth-like planet in the Solar System. It has basins, plains, and highland regions that are recognized as continents. It is tilted on its axis to a similar degree to Earth, which endows the planet with four seasons. Its day is 24 hours 37 minutes long. Like Earth, Mars has polar ice caps, and it retains some semblance of an atmosphere. But that's where the similarities end.

AN UNEARTHLY WORLD

The Martian atmosphere is thin—equivalent to an altitude of 140,000 feet (42,500 m) on Earth—and consists of 95 percent carbon dioxide. Although carbon dioxide is a greenhouse gas which helps keep a planet warm, Mars' atmosphere is so tenuous it retains little solar heat. As well, the planet's highly elliptical orbit accentuates seasonal differences. Consequently, the climate is extremely harsh, with temperatures ranging from −193°F (−125°C) in the polar winter to 62°F (17°C) in the southern summer. At some locations on Mars, it is cold enough for carbon dioxide to freeze out of the atmosphere as dry ice. Finally, its surface is dry and desolate, with no ecosystem or oceans.

A WATERY WORLD THAT RAN DRY

Four and a half billion years ago, Mars was a very different planet indeed. It had a thicker atmosphere and enough heat left over from its formation to melt water ice at the planet's surface. Judging from the many rivulets, channels, and canyons that have been seen by space probes, and by analysis

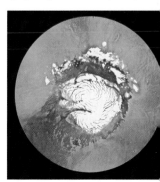

A WINDY PLACE

Although Mars' air is extremely thin, winds nonetheless can loft dust particles about one micron across. On Earth, dust particles of this size can be raised by a wind speed of only 20 miles per hour (32 km/h), but on Mars, winds need to reach at least 110 miles per hour (177 km/h).

findings pointed to thick deposits of sedimentary rocks, possibly laid down in now-vanished lakes or oceans.

DESERT STORMS

Mars consists mainly of silicate rock and it retains a modest metal core. The surface is a desert of rusty rocks and dust racked by intense surface winds that on occasion whip up immense dust storms. Through a telescope, these storms sometimes appear as localized yellow clouds that soon settle out.

At other times, particularly as the planet approaches perihelion, when it is closest to the Sun, the storms take on global proportions, engulfing the entire planet for months at a time. Such a storm occurred in 1971 at the beginning of the Mariner 9 mission, and again in 1977 during the Viking mission. The Mars Pathfinder spacecraft has also seen evidence of recent dust storms.

A Martian Lake?

The Martian climate must have been very different in the past. Spacecraft have discovered numerous channels cut by running water. A huge lake might once have filled the dark area on the left of this Viking image of Candor Chasma. Flowing water might also have had a hand in forming the deeply cut terrain on the right.

of the compositions of Martian meteorites found on Earth, abundant water must once have flowed across Mars' surface.

Where is the water now? Water still exists in the planet's polar caps and atmosphere. In 2000, planetary scientists analyzing images from NASA's Mars Global Surveyor spacecraft reported gullies in crater walls that suggest groundwater may lie at depths of 330 to 1,300 feet (100 to 400 m). Other spacecraft

Moons
Mars' two little moons, Deimos (left) and Phobos (right), were discovered in 1877 by the U.S. astronomer Asaph Hall.

Olympian Majesty
Olympus Mons is a shield volcano, similar in type to Hawaiian volcanoes but far larger at 340 miles across and 15 miles high (550 km by 25 km). Its eruptions and lava flows must have been correspondingly more extensive and massive.

A VOLCANO TO SURPASS ALL OTHERS

Mars has some spectacular features, one of the most prominent being Olympus Mons, perhaps the largest volcano in the Solar System. It rises 13 miles (21 km) above the surrounding plains on the planet's western hemisphere, and covers an area as large as Arizona. Three other large volcanoes lie to the southeast in a region of ancient volcanic activity called the Tharsis bulge, or ridge.

The Valles Marineris, just south of the equator, is also remarkable. This system of canyons up to 4 miles (7 km) deep forms an immense gash stretching some 2,500 miles (4,000 km) across the planet. Scientists think that activity in the Tharsis bulge broke open the crust and widened the canyon as ice washed out of the canyon walls.

TWO SMALL MOONS

Mars' two moons are tiny and may be asteroids captured by the planet's gravity. The larger, Phobos, is a potato-shaped body measuring 8.4 miles (13.5 km) across which orbits the planet every 7.7 hours. Gravity is so weak there that someone with an Earth weight of 150 pounds (68 kg) would weigh only 2 ounces (57 g). Even smaller is Deimos, lying two and a half times farther out than Phobos. This chunk of rock whizzes around Mars in just 1.26 days.

MARS	
Diameter	4,221 miles (6,794 km)
Mass	0.11 Earth masses
Rotation period	1.03 days
Inclination of equator to orbit	25.2 degrees
Mean orbital speed	15 miles per second (24.1 km/s)
Mean distance from the Sun	142 million miles (228 million km)

WHERE ARE THE MARTIANS?

In the early 1900s, the U.S. astronomer Percival Lowell founded an observatory for the specific purpose of finding life on Mars, which he firmly believed existed. He saw a system of irrigation canals lacing the planet—evidence for the engineering skills of an advanced civilization.

Lowell's ideas have long been disproved, but the possibility of life on Mars persists. We now have evidence of landforms caused by flowing water. Could some form of life have existed when Mars' climate was less severe? The 1976 Viking missions, however, found no traces of organic compounds in the soil. Whether life once existed on Mars—or if it still exists—is a question that only extensive exploration will settle.

Future Mars missions include two NASA rovers, planned for arrival in 2004, that are mobile versions of the one that performed so successfully during the 1997 Mars Pathfinder mission (see p. 60). These rovers will carry sophisticated instruments that will search for evidence of liquid water as well as study surface geology. Sample-return missions planned for the 2010s raise the tantalizing prospect of retrieving Martian surface material for study back on Earth.

Viking's Search for Life
The Viking 1 lander set down on Mars on July 20, 1976. This is what it saw—a desolate, rock-strewn landscape under a dirty pink sky. The lander's main mission was to look for traces of life. Using its robotic arms, it scooped up soil samples and chemically analyzed them for any organic substances. The results, and those of Viking 2 on the opposite side of the planet, were negative.

Return to Mars
Sojourner Rover ambled over the Martian surface in 1997 taking geological measurements. Similar rovers will one day return to Mars to look for life.

Asteroids and Meteoroids

Scattered throughout the Solar System are millions of asteroids and meteoroids, rocky bodies ranging in size from mammoth mountains to microscopic grains.

Eros
Eros is a piece of rock about 21 miles (34 km) long, irregular in shape and pockmarked with craters. It rotates once every 5¼ hours.

Asteroid Orbits
The elliptical orbits marked on the diagram belong to some of the near-Earth asteroids—ones that cross the path of Earth. The vast majority of asteroids, however, orbit in the main belt. The Trojans are two clusters of asteroids that lie in the orbit of Jupiter.

Asteroids, also known as the minor planets, are metallic, rocky objects without atmospheres. Astronomers believe asteroids consist of material that failed to form a planet-size body when the Solar System was taking shape less than five billion years ago, probably due to Jupiter's gravitational dominance. If such a planet *had* consolidated, it would have been very small—only 930 miles (1,500 km) across, or less than half the diameter of the Moon. More than 35,000 asteroids have well-surveyed orbits, but establishing exactly how many exist is probably impossible.

MAIN BELT ASTEROIDS

Tens of thousands of asteroids occupy a vast doughnut-shaped ring between Mars and Jupiter called the main asteroid belt. Asteroids were detected there in the 19th century, and the first four discovered are considered to be the principal minor planets. They are Ceres, the largest known at 567 miles (913 km) across; Pallas, at 325 miles (523 km), the second largest; Juno, the smallest of the four, with a diameter of 145 miles (234 km); and Vesta, 323 miles (520 km) across. Sixteen asteroids have diameters of 150 miles (240 km) or greater. Most asteroids in the main belt take between three and six years to complete a full circuit around the Sun.

Scientists classify asteroids by how much light they reflect and their chemical composition. Three major types stand out. The dark C-type or carbonaceous asteroids make up about three-quarters of all asteroids. They dominate the outer main belt and are rich in clays and water-bearing minerals. The S-type, about 17 percent of the whole, are commonest in the inner belt

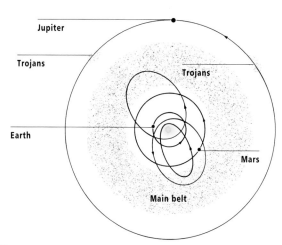

Jupiter

Trojans

Trojans

Earth

Mars

Main belt

nd have a rocky composition.
Dimmer than the S-types, the
M-type or metallic asteroids orbit
n the middle of the belt and are
argely nickel-iron.

TROJANS AND NEAR-
EARTH ASTEROIDS

Not all asteroids are found within
the main belt. Jupiter has two
clutches of asteroids traveling
with it in its orbit, one ahead
and one behind. They are known
as the Trojan asteroids, and are
traditionally named after figures
from the Trojan Wars.

Other asteroids have trajec-
tories that take them toward us.
An asteroid that comes within

121 million miles
(195 million km) of
the Sun is known
as a near-Earth
asteroid (NEA).
Some of these have
orbits that cross
Earth's, with a risk of
collision. There are about
500 known NEAs, but
astronomers think there may be
thousands more that are large
enough—0.6 miles (1 km) in
diameter—to cause devastation
if they hit Earth. Many scientists
believe that the impact of an
asteroid or comet was responsible
for wiping out the dinosaurs 65
million years ago.

NEAR
The Near-Earth Asteroid
Rendezvous (NEAR)
spacecraft was launched
to study the asteroid Eros.

Rocks in Space
The photographs at left
show, left to right, the
asteroids Mathilde, Gaspra,
and Ida, all to scale.
Mathilde is scarred by
more large impacts than
the other two, perhaps
reflecting a more violent
history. Ida and its moonlet
(below) have been color-
enhanced to reveal
mineralogical details.

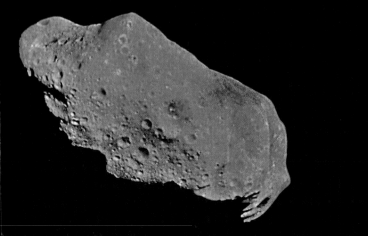

Touching Outer Space
At a museum display, schoolchildren reach out to touch a piece of outer space—a giant meteorite.

Meteor Storm
One of the biggest meteor storms on record occurred on November 17, 1833, when the Leonid meteor shower put on a priceless performance. All along North America's east coast stunned observers saw hundreds of meteors per minute. Estimates of the hourly rate range from 50,000 to 200,000.

METEOROIDS AND METEORITES

Earth experiences a continuous rain of interplanetary objects, known as meteoroids, which range in size from microscopic particles to small boulders. When they enter the atmosphere —visible as a streak of light, a "shooting star"—they are called meteors. Those that make it to the ground are meteorites.

Meteoroids are fragments of asteroids or comets. Debris from comets is fragile and burns up high in the atmosphere. But a piece of an asteroid is often larger and tougher; many survive to land as meteorites.

Stony meteorites consist of iron- and magnesium-bearing silicates and appear to be relatively unchanged since the Solar System formed. Iron meteorites, the second most common type, consist of 90 percent iron and 10 percent nickel with traces of silicates. These meteorites appear

to be more directly linked to asteroids that underwent a molten period, causing their heavier metals and lighter rocks to separate into different layers. The rarest meteorites are the stony-irons, intermediate between the other two.

Meteors appear between 50 and 75 miles (80 and 120 km) above Earth's surface and move at velocities between 25,000 and 160,000 miles per hour (11 and 72 km/s). Larger meteors often explode or fragment and leave a visible smoke trail, or train, that can be tens of miles long.

METEOR SHOWERS

On most moonless nights, you could expect to see around four or five meteors per hour. Meteor showers offer something much more spectacular.

When a comet comes out of the deep freeze of space and enters the inner Solar System, its surface is warmed by the Sun. Part of its icy crust suddenly turns

to vapor, which solar radiation blows off the comet's surface, creating a trail of ice, dust, and grit. When the Earth sweeps through this detritus, the particles burn up in our atmosphere, creating a meteor shower.

Meteor showers are named after the constellation or star nearest to the point in the sky from which they appear to come, a region called the radiant. The best viewing of a meteor shower occurs when the radiant is more than halfway up in the sky. City lights and the Moon's glare obscure the fainter members of a shower, as do hazy skies.

A meteor shower's activity is gauged by its zenithal hourly rate, or ZHR. This is the number of meteors an observer may see per hour in a very dark, clear sky when the radiant is overhead and the shower is at its peak. These ideal conditions are rarely met, so do not be surprised to see fewer meteors than the ZHR promises.

Space Rocks

Meteorites come in three varieties. As their name suggests, stony-iron meteorites (far left) are a mixture of stony material and iron. They make up just 1 percent of meteorites. Far more common are the stony meteorites (center). Iron meteorites (right) make up only 7 percent of all meteorites.

MAJOR ANNUAL METEOR SHOWERS

Name	Active period	Peak	Radiant	ZHR
Quadrantids	Jan 1–Jan 5	Jan 4	15h 20m +49°	120
Gamma Normids	Feb 25–Mar 22	Mar 13	16h 36m −51°	8
Lyrids	Apr 16–Apr 25	Apr 21	18h 04m +34°	15
Eta Aquarids	Apr 19–May 28	May 4	22h 32m −01°	20
Delta Aquarids	Jul 12–Aug 19	Jul 27	22h 36m −16°	20
Perseids	Jul 17–Aug 24	Aug 12	03h 04m +58°	100
Alpha Aurigids	Aug 24–Sep 5	Aug 31	05h 36m +42°	10
Draconids	Oct 6–Oct 10	Oct 8	17h 28m +54°	variable
Orionids	Oct 2–Nov 7	Oct 21	06h 20m +16°	20
Taurids	Oct 1–Nov 25	Nov 2	03h 28m +13°	10
Taurids	Oct 1–Nov 25	Nov 12	03h 52m +22°	10
Leonids	Nov 14–Nov 21	Nov 17	10h 12m +22°	100
Puppid-Velids	Dec 1–Dec 15	Dec 6	08h 12m −45°	10
Geminids	Dec 7–Dec 17	Dec 14	07h 28m +33°	100
Ursids	Dec 17–Dec 26	Dec 22	14h 28m +76°	20

A meteor shower's zenithal hourly rate, or ZHR, represents the number of meteors you might see per hour given prime observing conditions during the shower's maximum. The radiant is the region in the sky from which the meteors will appear to streak. The sky coordinates are given in right ascension (hours and minutes) and declination (degrees). For more information on observing meteors, see p. 254.

Leonid Meteors

The brilliant yellow streaks in this false-color image belong to the Leonid meteor shower, which reaches a peak of activity in mid-November each year. The shower appears to come from a region of sky in the constellation Leo, hence its name.

Battle Scars of the Planets and Moons

In July 1994, the fragmented Comet Shoemaker-Levy 9 collided with Jupiter, releasing energy equivalent to millions of megatons of TNT. Upon entering the gas giant's upper atmosphere, each fragment exploded in a fireball which created huge plumes of dust, some of them over three times Earth's diameter.

Such an occurrence is a dramatic reminder of the vulnerabilty of all planets and their moons to collisions with comets and asteroids. Numerous impact craters have been mapped on Mercury, Venus, the Moon, and Mars. Fewer craters are visible on Earth only because geological activity has remade the planet's surface many times over.

Direct Hit on Jupiter
The Hubble Space Telescope saw fragments of comet Shoemaker-Levy (lower picture) collide with the giant of the Solar System— Jupiter. The bright patches on the lower part of the Jovian disk (upper picture) show four of the impact sites

Impact at Gosse Bluff
One of the most recognizable impact sites on Earth is Gosse Bluff in the Australian outback. About 2 miles (4 km) wide, it is the remnant of a much larger crater formed 130 million years ago by an asteroid or comet

The Battered Moon
The huge number of lunar craters is reminder that the Moon's early history was very violent. The scars remain because processes to recycle the crust, such as weathering, do not take place on the Moon

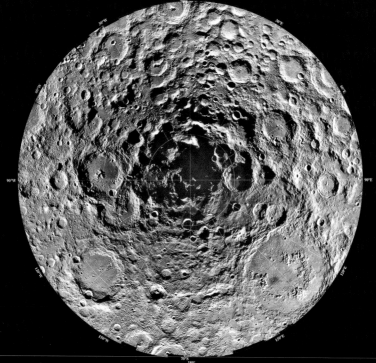

Jupiter

Jupiter was named after the chief god in Roman mythology, and rightly so. Its enormous size and great mass make it the undisputed senior member of the Solar System.

Colorful Giant
Jupiter presents a complex and colorful face of dark belts, light zones, and cloud structures such as the Great Red Spot. The Jovian day spins the belts and zones into streaks and disturbances, driven by winds that blow in alternating directions parallel with the equator.

The statistics speak eloquently of Jupiter's great size: its equatorial diameter of 88,846 miles (142,984 km) is 11 times that of Earth; it is more than 317 times as massive as Earth and twice as massive as all the other planets combined; and its gravitational pull is second only to the Sun's. Had Jupiter been 80 times more massive, pressure from its outer layers could have raised interior temperatures enough to trigger nuclear reactions, thereby making it a low-mass star.

DISCOVERING JUPITER'S SECRETS

Jupiter is on average about 485 million miles (780 million km) from us, making the gas giant very difficult to observe in detail by telescope. Astronomers had to wait until the space age to see

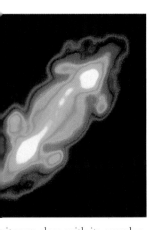

ɔiter up close, with its complex
nosphere, diverse retinue of
ɔons, and faint ring system.
Flybys by Pioneer 10 in 1973
d a year later by Pioneer 11
gan the process. Voyager 1
ched Jupiter in March 1979,
th Voyager 2 following a few
ɔnths later. Both spacecraft
ide detailed images of the
inet's atmospheric features
d its four largest satellites.
Galileo, which arrived at the
ʌian system in December 1995,
s provided many eye-opening
ages of Jupiter and the surface
itures of its satellites. It also gave
e first directly obtained data of
e physical and chemical nature
the planet's gaseous upper
ers by deploying a probe into
ɔiter's atmosphere.

COLORFUL CLOUDS AMID
A WHIRLPOOL OF WIND

Crossing Jupiter's disk are bright
and dark streaks, known to
astronomers as zones (the light-
colored bands) and belts (the
dark ones). They are long-lasting
features, having remained at
the same latitudes for at least a
hundred years. Earth- and space-
based observations have shown
that they are created by jets of
wind circulating in alternating
directions. The Galileo probe
discovered that the winds blow
at 400 miles per hour (643 km/h)
and may extend thousands of
miles below the cloud tops.
These wild winds may be driven
by convection currents produced
by an internal energy source—
Jupiter radiates about twice as
much energy as it receives from
the Sun because it is still cooling
off from its formation.

The clouds' markings are the
result of chemical reactions and
are also related to the clouds'
altitude. Blue clouds lie deepest
within the atmosphere, while
brown and white clouds occur
higher up. Rust-colored clouds
are highest of all, and may be
colored by traces of phosphorus
brought up by convection from
Jupiter's interior.

Magnetic Field

Jupiter has a powerful
magnetic field generated
by convection currents in
its hot interior. This radio
image shows the planet
in the center flanked with
bright belts of charged
particles trapped by its
magnetic field.

Galileo

In 1995, the spacecraft
Galileo arrived to orbit
Jupiter, and shot a
probe into the planet's
atmosphere. The probe
radioed back data for more
than an hour before being
destroyed by high pressure
and temperature. The
orbiter then studied the
planet and its
family of
satellites for
several
years.

PITER	
ameter	88,730 miles (142,796 km)
iss	317.8 Earth masses
tation period	0.41 days
lination of equator to orbit	3.1 degrees
ean orbital speed	8 miles per second (13 km/s)
ean distance from the Sun	485 million miles (778 million km)

High Clouds
An infrared image of the Great Red Spot shows that the highest clouds (colored pink) are in the center.

Internal Structure
Jupiter consists almost entirely of hydrogen and helium. A core of molten rock, perhaps several times Earth's mass, is thought to lie at its center.

Molecular liquid hydrogen

Rocky core

Metallic liquid hydrogen

Gaseous hydrogen

THE GREAT RED SPOT

Of all the cloud formations on Jupiter, the most famous is the Great Red Spot, which has been observed for more than 300 years. It varies in size—and color intensity—and at its largest is 25,000 miles (40,000 km) long and 8,700 miles (14,000 km) wide. Infrared observations from the Pioneer and Voyager probes show that it is a colossal eddy.

Similar high-pressure structures appear elsewhere on Jupiter, as well as on Saturn and Neptune, but none matches the Great Red Spot in size. No one knows what sustains these systems over such long periods.

A CRUSHING WORLD BENEATH THE CLOUDS

As a gas giant, Jupiter's structure and composition are radically different to those of the terrestrial planets. The atmosphere consists of 86 percent hydrogen and 14 percent helium with traces of methane, water, and ammonia. This brew is thought to be much like the composition of the solar nebula, from which the Solar System formed.

Just beneath the cloud tops is a layer of hydrogen gas. Some 600 miles (1,000 km) below this the hydrogen becomes liquefied a point that marks the boundary

HE GREAT COMET-ATTRACTOR
piter's powerful gravity makes
a magnet for passing comets.
July 1994, the fragments of
omet Shoemaker-Levy slammed
to Jupiter, creating spectacular
xplosions and dust plumes in
e planet's atmosphere that
ere clearly visible in amateur
lescopes. The dark clouds from
e collisions remained visible
r nearly a year by the Hubble
ace Telescope.

etween the planet's outer atmos-
here and its interior. At a depth
f between 16,000 and 19,000
iles (25,000 to 30,000 km),
nder an estimated 4 million
ars of pressure, the hydrogen
ecomes metallic and can con-
ct electricity. It is thought that
is is what generates Jupiter's
owerful magnetic field, 20,000
mes stronger than Earth's.

In the upper atmosphere,
mperatures are about −162°F
108°C). Temperatures increase
40,000°F (22,000°C) at the
anet's center, which may con-
in a core of rocky material, 10
20 times the mass of Earth.

F RINGS AND MOONS
yager 2 confirmed a finding
Pioneer 11 that Jupiter is
ncircled by a faint system
rings

made up of small rocky grains.
Unlike Saturn's rings, the Jovian
rings contain little or no ice.

The four largest of Jupiter's
39 known moons—Callisto,
Ganymede, Europa, and Io—were
discovered in 1610 by Galileo and
are referred to as the Galilean
satellites. Ganymede and Callisto
are larger than Mercury and
Pluto, while Europa and Io are
similar in size to Earth's Moon.
Each of the four has fascinating
features: astronomers suspect that
beneath Europa's frozen surface
lies an ocean; Io is the most
volcanically active world in the
Solar System, with hundreds of
volcanoes and vents ejecting
massive amounts of material;
Callisto is densely cratered, with
one crater measuring 900 miles
(1,500 km) in diameter; and
Ganymede, the Solar System's
largest moon, has a complex
surface and perhaps a sub-
surface ocean like Europa's.

Jupiter's other moons
are much smaller. Several
outer satellites are thought
to be captured asteroids—
one of the smallest of these,
Erinome S/2000 J 4, is
only 2 miles (3.2 km)
in diameter.

Galilean Satellites
The Galilean satellites are
Europa, Io, Callisto, and
Ganymede. Europa's icy
surface resembles a planet-
wide skating rink. Io, the
closest of the four to
Jupiter, is so affected by
the planet's tidal forces
that its surface is being
constantly repaved by volcanic eruptions. The outer
two moons have older, icy
surfaces, both scarred with
craters, but differing in
other features and telling
of different histories.

Ganymede

Callisto

Io

ropa

Saturn

Saturn shares much in common with its neighbor, Jupiter, but a spectacular ring system sets it apart, making it one of the most instantly recognizable worlds in the Solar System.

In many respects, Saturn is Jupiter's little brother. It has a similar composition to Jupiter—about 96 percent hydrogen and 3 percent helium. This gaseous makeup gives the two planets low densities compared to the terrestrial planets, but Saturn is much less dense than Jupiter—in fact, it has the lowest mean density of all planets, 0.7 times that of water.

The structure is similar, too. A rocky core is embedded within an outer core of water, methane, and ammonia. Above this is a layer of liquid metallic hydrogen, some 13,000 miles (21,000 km) deep, over which lies a stratum of molecular hydrogen.

FIERCE WINDS AND GREAT STORMS

Also like Jupiter, Saturn radiates more energy than it receives from the Sun. The continuous release of interior heat is likely responsible for generating convection currents in the atmosphere, which, in turn, gives rise to Saturn's high-altitude clouds. The equatorial jet winds at the cloud tops move faster than they do on Jupiter, about 1,100 feet (500 m) per second, which amounts to two-thirds the speed of sound in the atmosphere there.

Bands and Rings
Voyager 2 snapped this close-up of Saturn on its flyby in 1981. The false colors emphasize the planet's banded clouds and the divisions of the rings.

SATURN	
Diameter	74,900 miles (120,500 km)
Mass	95.2 Earth masses
Rotation period	0.43 days
Inclination of equator to orbit	26.7 degrees
Mean orbital speed	6 miles per second (9.6 km/s)
Mean distance from the Sun	890 million miles (1.4 billion km)

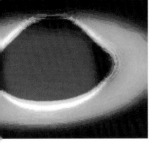

Saturn orbits much farther om the Sun than Jupiter does, it is colder and therefore periences less "weather" and splays fewer cloud features an Jupiter. Nevertheless, larger escopes can make out bands clouds, and Voyager probes vealed a variety of cloud struc- res: long-lived white ovals, ains of vortices, and cloud con- nsations. In 1990, the Hubble ace Telescope recorded an ormous storm that erupted near the planet's equator, an event that happens roughly every 30 years during summer in the planet's northern hemisphere.

Saturn rotates rapidly (10 hours and 14 minutes at the equator), which is largely responsible for the planet's flattened, or oblate, appearance. The other gas giants also exhibit some oblateness, but not to the same extent as Saturn.

THE GLORY OF THE RINGS

Saturn's rings are immense—and very thin. About 170,000 miles (274,000 km) in diameter with a width of some 39,000 miles (62,000 km), they are less than a mile and a half (2 km) thick. As seen from Earth by a small telescope, they look like a solid disk encircling the planet. In fact, they consist of many smaller

Picturing the Rings

This false-color image (above) spans Saturn's rings from the inner C ring (bluish color at right), through the B ring (warm tints merging into greens and blues), out to the Cassini Division (dark with bright streaks). Outside the Cassini Division lies the A ring. An infrared image made from Earth (left) highlights the differences in composition between the ring system and the planet's disk—the mostly icy rings appear blue, Saturn's methane-rich atmosphere looks red.

Origin of the Rings

How did Saturn's rings form? One theory suggests that an object collided with an icy moon of Saturn, shattering it. In orbit about the planet, the debris was ground to small chunks by collisions. Eventually, the chunks spread out to form rings, gravitationally kept in place by small moons.

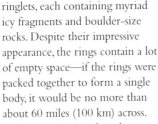

ringlets, each containing myriad icy fragments and boulder-size rocks. Despite their impressive appearance, the rings contain a lot of empty space—if the rings were packed together to form a single body, it would be no more than about 60 miles (100 km) across.

Three main rings have been recognized for many years: two conspicuous ones, called A and B, and a fainter one, called C, or the crepe ring. A gap, called the Cassini Division, separates rings A and B—"gap" is actually misleading because this region contains four narrow ringlets, each about 300 miles (500 km) across. The narrower cleft located in the outer part of the A ring is called the Encke gap.

In the 1970s and 80s, Voyager and Pioneer 11 images revealed four additional faint rings. The innermost D ring consists of a few dusty and widely spaced bands. The outermost E and G rings are, in essence, slight concentrations of orbital debris and are

very faint. Pioneer 11 detected the unusual F ring lying 2,200 miles (3,500 km) outside the A ring. This ring appears smooth in places, but also has a set of knotted and braided strands, shapes that arise partly from the gravitational effects of the "shepherd" satellites, Pandora and Prometheus.

The two Voyager probes also confirmed the existence of dark, shadowy, spoke-like structures that appear across the rings sporadically. Astronomers think they consist of microscopic grains and may be shaped by an interaction between the particles and Saturn's magnetic field.

REASONS FOR THE RINGS
We are still not sure why Saturn has rings, or why any other planet has them, for that matter. One theory says that the rings are the remains of a moon that either strayed too close to Saturn and was pulled apart by the planet's gravity, or was smashed to pieces by an asteroid or comet. Another theory holds that the rings are made of particles that were unable to concentrate into a moon, again because of the gravitational influence of the planet.

Iapetus

Enceladus

Mimas

tan

LITTLE WORLDS OF ROCK AND ICE

Saturn has at least 30 known satellites, ranging in ze from about 2 miles (3 km) ross to Titan, larger than ercury and the second-largest oon in the Solar System. Titan unique among moons in having hick atmosphere—mostly trogen with trace elements, hich include methane, ethane, etylene, carbon dioxide, and rbon monoxide. The moon

Iapetus is known for its peculiar dark and bright hemispheres. The dark material is thought to be carbon-based, but astronomers do not know if it came from within Iapetus or was deposited from space. Enceladus consists largely of water-ice and appears to be the most geologically active of Saturn's moons. Mimas has been pockmarked by impacts— a particularly large crater is the result of a collision that must have almost torn the moon apart.

Saturn's Brood

Saturn's moons have an astonishing variety of features, ranging from the thick, smoglike atmosphere of Titan to the enormous "bull's-eye" crater on Mimas. The composite image of photos taken by Voyager 1 (below) shows Saturn with six of its brood, from left to right: Titan, Mimas, Tethys, Dione (at front), Enceladus, and Rhea (at top right).

Uranus

Uranus is an enigmatic blue-green world which has only recently begun to give up its secrets. Third largest of the gas giants, it is nearly four times the size of Earth.

Uranus' World
The Hubble Space Telescope recorded this infrared image of Uranus, which displays a wealth of detail, including the rings, six of the moons, and various cloud features. Green and blue parts of the planet's disk show that the atmosphere is clear; yellow and gray colors indicate sunlight reflecting from a higher haze or cloud layer; and high clouds show up as orange and red.

Being so remote, some 1.8 billion miles (2.9 billion km) from the Sun, little was known about Uranus until relatively recent times. Most of our knowledge of the planet came from a single spacecraft—Voyager 2—which sped by Uranus at close range in 1986.

A PLANET ON ITS SIDE

Uranus has its rotational axis oriented close to its orbital plane; in effect, it orbits on its side. This has strange consequences. During its 84-year orbital period, each pole spends 42 years in darkness followed by 42 years in direct sunlight. You would expect the extra solar energy received at the poles to translate into higher temperatures there, but this is not the case. Temperatures are still higher at the equator because the planet's dense atmosphere is a good insulator, and it takes the atmosphere longer than Uranus' orbital period to heat up or cool off.

BENEATH THE BLUE-GREEN CLOUDS

Like the other gas giants, Uranus is thought to have a small rocky core enveloped in a thick layer of liquid ice and rock, but unlike Jupiter and Saturn it lacks a layer of metallic hydrogen. The planet's outer layer consists largely of hydrogen and helium, though in

URANUS	
Diameter	31,800 miles (51,100 km)
Mass	14.5 Earth masses
Rotation period	0.72 days, retrograde
Inclination of equator to orbit	97.9 degrees
Mean orbital speed	4.2 miles per second (6.8 km/s)
Mean distance from the Sun	1.8 billion miles (2.9 billion km)

uch smaller proportions than in
ne two larger gas giants. At high
titudes, the atmosphere is so
old that methane and ammonia
ondense into clouds of ice
rystals. The methane is largely
sponsible for the planet's blue-
reen color.

ARROW RINGS AND
LARGE FAMILY

narrow ring system was
etected by Earth-based astron-
mers in 1977, and Voyager 2
onfirmed the discovery in 1986.
he 11 known rings are dark, like
piter's, but contain larger par-
cles, like Saturn's. The outer-
ost, or Epsilon, ring is the
rightest, although it comes
owhere near matching the
rilliant colors of Saturn's rings.

Up until Voyager's visit to
ranus in 1986, we knew of just
moons; Voyager spied 10 more,
d subsequent discoveries have
ded 6—a total of 21. Almost all
e named after characters from
akespeare's plays.

Most of the moons are tiny
orlds of rock and ice only 40 to
miles (25 to 100 km) across.
ve primary satellites, however,
e visible from Earth, and these
nge in diameter from about
000 miles (1,600 km) to
0 miles (485 km).

Titania is the largest—about
lf the size of our Moon—but
is a much smaller one that has
cited most interest. Voyager
ages of Miranda show three
tally distinct types of terrain
xtaposed to each other: cratered
ins, bright cliffs and scarps, and
oved regions extending nearly

200 miles (300 km) across the
surface. Perhaps Miranda was
shattered in a collision and the
pieces subsequently reassembled,
but no one knows for sure.

Miranda
Miranda is a moon like no
other. Consisting largely of
water-ice and coated with
dust, its surface appears to
have been thrown together
on a whim. The jumble of
terrains includes rolling,
cratered regions, fractures,
and systems of parallel
ridges and troughs called
coronae. Scientists are still
working out how Miranda
came to look like this.

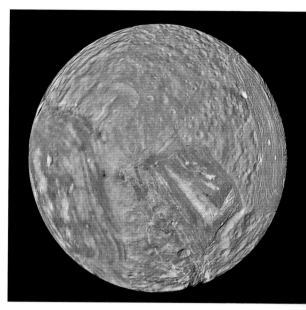

Neptune

Neptune is the smallest and most distant of the gas giants. It is also one of the most interesting, with a surprising amount of "weather" and a moon whose vents erupt nitrogen not lava.

Neptunian Cavalcade
Neptune is the smallest of the gas giants, but it still has a mass equal to about 17 Earths. An ever-changing cavalcade of atmospheric bands, bright ammonia-ice clouds, and dark, stormy regions march around its bluish disk. The faint ring system is not visible in this Voyager photo.

When Voyager 2 approached the Solar System's eighth planet on August 25, 1989, scientists were uncertain what it would find. Would Neptune appear as featureless and static as Uranus? They soon got their answer as the first images were transmitted to Earth. Instead of a bland, cueball-like planet, scientists saw a blue disk punctuated with a great dark oval, dark bands, and numerous strata of wispy white clouds. None of these features had ever been seen before.

VIOLENT STORMS ON NEPTUNE'S SPHERE

Neptune's composition and structure resemble those of its larger neighbor, Uranus. An iron-silicon core, overlain with a mixture of various ices and rock, is thought to lie at its center. The atmosphere is mostly hydrogen and helium gas with traces of methane. Temperatures in the upper part of the atmosphere, as on Uranus, are so low that methane freezes. Unlike Uranus, however, Neptune radiates more than twice as much energy as it receives from the Sun, evidence of an internal source of heat. It is this energy that powers Neptune's violent winds and storms.

When Voyager 2 took a peek at Neptune in 1989, the planet's southern hemisphere was dominated by an oval-shaped high-pressure region nearly 6,000 mil in length (10,000 km) and with supersonic winds flowing at its periphery. Below this Great Dar Spot was a smaller storm dubbe the Small Dark Spot. Voyager 2 also imaged a small, irregular white cloud which was christened Scooter because of its rap motion. This cloud, astronomers think, may be an upwelling plu from the lower atmosphere.

In 1995, observations of Neptune by the Hubble Space Telescope confirmed the dynam nature of the atmosphere. There was no trace of either dark spot and new features had emerged. The spots either faded, as Jupite Great Red Spot does from time to time, or dissipated altogether

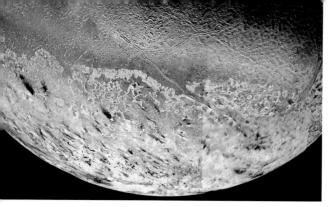

Moons Great and Small

Neptune's largest moon, Triton, is a complex world. Relatively few impact craters are visible in this Voyager view (left). Instead, it has a fairly young-looking surface with a rough texture that bears a strong resemblance to the skin of a cantaloupe. Visible at the polar cap are dark streaks deposited by geysers. Triton is a giant compared to little Nereid (left), an indication that these moons had very different origins.

IM RINGS AND AN INTRIGUING MOON

eptune has a tenuous ring stem containing some bright umps distributed along the ng's arc. There are only four nown rings, and because they e so distant and faint, precious tle is known about them.

Of the eight satellites, Triton the largest, with a diameter 1,700 miles (2,700 km), about o-thirds the size of the Moon d a little larger than Pluto. With inimum temperatures of −400°F 240°C), Triton's surface is as ld as Pluto's. Its composition rock and ice is also thought resemble Pluto's.

Triton's surface features range m the run-of-the-mill (impact aters) to the extraordinary range dimpled areas of pits and pressions). In places, nitrogen s shoots upward from vents 5 miles (8 km) into the thin

atmosphere of nitrogen vapor, before wafting downwind to form dark plumes on the ground.

Triton orbits Neptune in a retrograde manner, that is, in the opposite direction to the planet's rotation. This motion could mean that long ago Triton was snared by Neptune's gravity.

Another moon, Nereid, about 200 miles (320 km) across, swings around Neptune in a long ellipse. So eccentric is its orbit that its distance from Neptune varies by more than 5 million miles (8 million km).

Team Effort

Neptune was discovered on September 23, 1846, by two German astronomers, Johann Galle (below) and Heinrich d'Arrest, but only after its position had been mathematically predicted first by John Couch Adams in England and independently by Urbain Le Verrier in France.

PTUNE	
ameter	30,800 miles (49,500 km)
ss	17.2 Earth masses
tation period	0.67 days
lination of equator to orbit	29.6 degrees
an orbital speed	3.3 miles per second (5.4 km/s)
an distance from the Sun	2.8 billion miles (4.5 billion km)

Volcanoes Reveal the Fury Within

A ctive volcanoes are rare in the Solar System—Earth and Jupiter's moon Io are the only known instances. (Venus is suspected also.) But evidence of past activity is widespread. Mars, Mercury, and the Moon all once had lava eruptions. Jupiter's moons Europa and Ganymede, Saturn's Enceladus and Tethys, the Uranian moons Ariel, Miranda, and Titania, and Neptune's Triton have all undergone volcanic activity. (Triton's geysers were active when Voyager flew past in 1989.) These moons display what scientists call "cryovolcanism." In this process, the lava is a slushy mixture of water and other materials, such as dust, ammonia, or methane.

Eruption or

Jupiter's moon Io shoots a plume of mol
sulfur into space. Eruptions like this, wh
occur constantly, have ensured that
surface is the youngest in the Solar Syt

Martian Gi

Mars has the Solar System's largest kno
volcano—Olympus Mons, 15 miles (25 k
high and 340 miles (550 km) acr

River of Ro

Eruptions on Earth occur when molten ro
or magma, from the interior forces its v
up through a weak zone in the crust. Wh
it does, the magma may flow like wa

Venusian Volcanoes

The tens of thousands of volcanoes on Venus include nearly 500 that are larger than about 12 miles (20 km) in diameter. Two of them are Sapas Mons (center foreground) and Maat Mons (background).

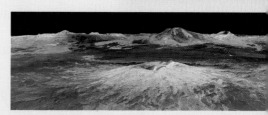

Pluto

Pluto is a maverick. It is tiny and has a tilted and markedly eccentric orbit. Definitely not a gas giant, neither is it a terrestrial world. Where does Pluto fit in?

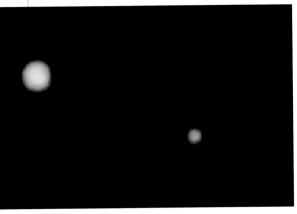

Far-distant Worlds
Pluto and its moon Charon are so far away from us that this Hubble Space Telescope image is our best view of these worlds so far.

Pluto's Orbit
Pluto's highly elliptical orbit is tilted relative to the orbits of all the other planets, and at times brings the planet closer to the Sun than Neptune.

Pluto lies in the dark, cold hinterlands of planetary space. It takes 248 years for Pluto to orbit the Sun, 2.8 billion miles (4.5 billion km) away at closest approach. At the moment it is the farthest planet from the Sun, but because Pluto's orbit is so elongated it is sometimes nearer to the Sun than Neptune, as it was from January 1979 until February 1999. Pluto's orbital path is also markedly tilted from the plane in which the other planets orbit. One final orbital oddity: Pluto orbits on its side, like Uranus.

PUTTING TOGETHER WHAT LITTLE WE KNOW
Pluto is the only planet that has not been visited by a spacecraft, making our knowledge of it relatively sketchy. We do know that it has a diameter of only 1,430 miles (2,300 km), making it the smallest planet—smaller even than our Moon. Pluto's mass is only one-fifth that of Earth with a rock-ice composition similar to Neptune's moon Triton. Observations in 1978 of Pluto and its moon Charon showed that the polar regions of these bodies are relatively brighter than their equatorial regions—but as yet we can only guess at the reasons why.

Pluto's atmosphere may contain mostly nitrogen with some carbon monoxide and methane. It is extremely tenuous, however, and

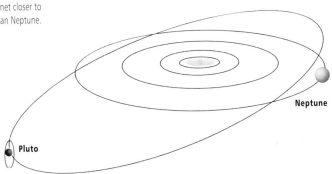

Neptune

Pluto

CLYDE TOMBAUGH

In 1929, Lowell Observatory in Arizona hired Clyde Tombaugh, an amateur astronomer, to look for a new planet believed to lie near Neptune. After a painstaking search, in which he compared thousands of photographic images, Tombaugh found what he was looking for on February 18, 1930.

...ay exist as a gas only when ...uto is closest to the Sun. The ...anet's surface temperature varies ...tween about −390° and −346°F ...230° and −210°C).

DOUBLE PLANET?

...uto has a close relationship to ... single moon, Charon—so ...ose, in fact, that some astron-...ners consider Pluto and Charon ... be a double-planet system. For ...e thing, Charon is very large

in proportion to the size of its host planet—its diameter of 745 miles (1,200 km) is slightly less than half that of Pluto's. And it is also very close to its planet, orbiting every 6.4 days at a mean distance of only 12,500 miles (20,000 km).

Additionally, the masses of these two bodies are so close that they spiral around a shared center of gravity that lies outside Pluto. In a normal planet–satellite system, such as Earth and the Moon, the center of gravity is near the planet's center.

COMETS GIVE CLUES TO PLUTO'S ORIGINS

Pluto's unique features have led some astronomers to theorize that it wandered in from the Kuiper Belt, a disk-shaped region lying beyond the zone of the planets. The bodies found there are icy planetisimals—comets without tails. These never accreted into larger objects like Neptune because they orbit too slowly and there are not enough of them to make planet-building collisions likely. Pluto may be the largest example of this group.

Pluto

Pluto and Charon
Charons's diameter is about half Pluto's, one reason why the two are considered by some to be a "planetary binary," or double planet. For a world on the dark edge of the Solar System, Pluto is aptly named: in Greek myth, Pluto is the god of the Underworld; Charon was the boatman who brought departed souls there across the River Styx.

Charon

...UTO	
...ameter	1,430 miles (2,300 km)
...ass	0.0022 Earth mass
...tation period	6.4 days, retrograde
...clination of equator to orbit	123 degrees
...ean orbital speed	3 miles per second (4.7 km/s)
...ean distance from the Sun	3.7 billion miles (5.9 billion km)

OUR SOLAR SYSTEM

Comets

A comet leads a double-life: for much of the time a dull mixture of ice and dust, it transforms itself as it approaches the Sun and turns on a magnificent spectacle .

Comet Halley
In 1705, Edmond Halley discovered that one comet kept returning about every 76 years. When it came back on schedule after Halley's death, it was named after him. The comet's last visit was in 1986; its next is in 2061.

Comets are often referred to as "dirty snowballs" because they consist of a mixture of ices (both water and frozen gases) and dust. Scientists think that these bodies are made up of material left over from the formation of the Solar System almost five billion years ago.

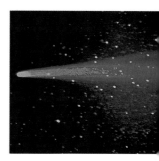

ICY VISITORS FROM BEYOND THE SOLAR SYSTEM

Astronomers believe that most comets reside in a vast sphere-shaped reservoir surrounding the Solar System called the Oort Cloud. The center of this region is a thousand times more distant than Pluto, about 50,000 astronomical units from Earth. Here, the gravity of passing stars can jolt comets on paths toward the inner Solar System.

A second, nearer source of comets is the Kuiper Belt, which is believed to be disk-shaped and lies just beyond the orbit of Pluto. Several hundred objects are known to orbit there; thousands more are suspected. These bodies

are called Trans-Neptunian Objects, and they include some that are up to 550 miles (900 km) in diameter.

ONE HEAD, TWO TAILS

As the comet approaches the Sun and inner Solar System, it feels an increase in heat. This vaporizes its long-frozen outer icy layers, which blow off a shell of gas and dust, called the coma, or head. It contains gas, dust, water, carbon dioxide, and other substances. As this material streams out of the coma, it creates the comet's most spectacular feature, its tail.

In fact, there are two tails. The dust tail is the curved, fan-shaped extension of smoke-size dust particles from the nucleus, and it can stretch between 5 and 60 million miles (10 and 100 million km). The ion, or plasma, tail is a bluish stream of ionized molecules lying adjacent to the dust tail. Often straight in appearance, it can extend for some 5 million miles (10 million km) or more. The gas and dust in the comet's tails are

Comet Structure
A typical comet has a tiny nucleus surrounded by a gaseous coma. Dust and ionized particles stream from the coma in two separate tails.

Ion tail

Coma

Dust tail

Nucleus

168

rced away from the Sun by
diation pressure and the
lar wind.

OMET ORBITS

omets travel about the Sun in
ongated, eccentric, orbits.
omets with orbital periods of
ss than 200 years are called
ort-period comets. Their orbits
e completely, or almost com-
etely, among the planets' orbits;
ey are probably gravitational
ptives of the giant planets.
omet Halley, with its 76-year
eriod, is one such example.

Comets with periods greater
an 200 years are called long-
eriod comets. They have para-
olic-shaped (open-ended) orbital

paths and approach the Sun from
all angles, indicating that they do
not come from any specific
direction in space.

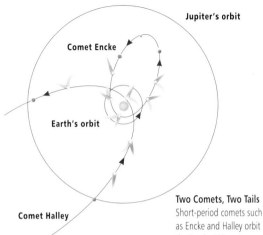

Jupiter's orbit

Comet Encke

Earth's orbit

Comet Halley

Two Comets, Two Tails
Short-period comets such
as Encke and Halley orbit
the Sun in 200 years or
less and travel among the
planets. Two tails stream
away from a comet when
it approaches the Sun—a
straight (blue) ion tail and
a curving (yellow) dust tail.

Comet Hale-Bopp
Comet Hale-Bopp was an
impressive sight in early
1997. This long-period
comet will next visit us
in about 2,400 years.

HE ORIGINS OF LIFE?

omets hitting Earth may have provided the building blocks for life on our
anet—carbon, hydrogen, oxygen, and nitrogen. Observation of Comet
alley in 1986 determined the presence of these materials in almost identical
roportions as exist on Earth.

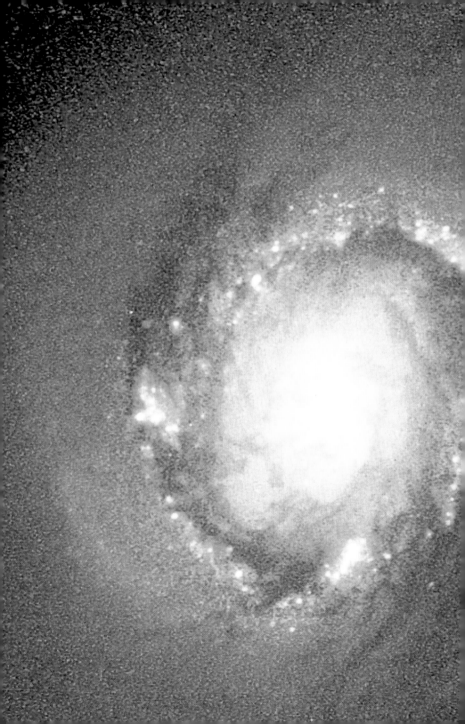

Stars and
Galaxies

Apart from the Moon and the planets, every fixed point of light we can see in the sky is a star, belonging to our Milky Way Galaxy. Space is filled with myriad galaxies, home to billions of stars. And stars and galaxies are only part of the picture: there are wispy nebulas, exotic black holes, enormous superclusters, and much more.

"Stars scribble in our eyes the frosty sagas,
The gleaming cantos of unvanquished space."

Hart Crane
U.S. poet, 1899–1932

Evolving Stars

Compared to our ever-changing world, the stars in the sky seem like models of stability and permanence. In truth, however, they have very definite lifespans and are slowly but surely evolving from birth toward death.

A star is a large sphere of hydrogen and helium, with a smattering of other elements, all in gaseous form. Nuclear fusion reactions in its core create enormous amounts of energy, including light and heat. No two stars are alike—brightness, surface temperature, and mass vary from one to the other.

Star Sizes
Stars range greatly in size. The Sun, a dwarf, is shown here as a yellow globe. The red giant behind it is a hundred times larger. By contrast, a white dwarf would be less than one-fiftieth the size illustrated. A neutron star might be a thousand times smaller again, portrayed here by a grossly oversized dot.

A STAR IN THE MAKING

Imagine a dense cloud of dust and hydrogen gas drifting through space. This giant molecular cloud may be disturbed by shockwaves from a nearby supernova or by gravitational perturbations from inside its own galaxy. Changes begin to take place: the cloud gets denser in places, and these regions begin to fragment and collapse under their own gravity. As the fragments continue collapsing, each forms a hot core, known as a protostar. The smaller a collapsing protostar becomes, the hotter it gets, until the core becomes hot enough to trigger nuclear fusion reactions at the core. Thereafter, radiation pressure halts the contraction. A star begins its life.

GROUPING THE STARS

Starlight gives clues to a star's surface temperature. Knowing that color intensity is a measure of surface temperature, astronomers use a spectroscope to split light from a star into a spectrum of colors. A star's color, temperature, and luminosity (absolute magnitude) are all interrelated. To see these relationships clearly, temperature can be plotted against luminosity. The result: the Hertzsprung-Russell (H-R) diagram (see p. 173), which captures a snapshot of stars evolving.

On the H-R diagram, stars fall into definite regions. The most conspicuous is a slightly curved band running from extremely bright, hot stars in the upper left corner to faint, cool stars in the lower right. This strip is called the main sequence. It contains 90 percent of all known stars and represents where a star spends most of its life. Giant and supergiant stars—bright but cool—occupy the upper right quadrant of the diagram, while white

DIAGRAM DUO

The astronomers who gave their n to the H–R diagram are the Dane Hertzsprung and the American He Norris Russell, who independently covered the temperature–luminos correlation in the early 1900s.

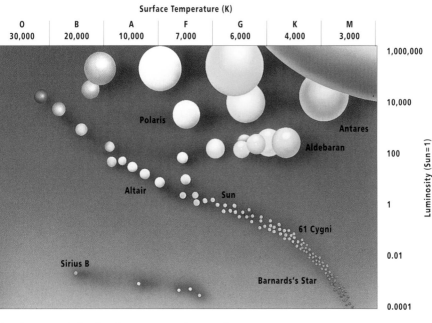

Surface Temperature (K)

O	B	A	F	G	K	M
30,000	20,000	10,000	7,000	6,000	4,000	3,000

Polaris

Antares

Aldebaran

Altair

Sun

61 Cygni

Sirius B

Barnards's Star

Luminosity (Sun=1)

1,000,000

10,000

100

1

0.01

0.0001

lwarfs, extremely hot but faint,
ie at lower left.

Stars are grouped into seven
main spectral classes in order of
decreasing temperature: O, B, A, F,
G, K, and M. O stars are blue and

can have surface temperatures
as high as 72,000°F (40,000°C),
while M stars are red and have the
lowest temperatures, about 6400°F
(3500°C). Each spectral class is
further divided into 10 subclasses,

H–R Diagram
The Hertzsprung–Russell
diagram presents the basic
groups of stars, together
with an indication of their
color and relative size. The
relative numbers of stars
in each portion of the
diagram are not correctly
presented. Most stars lie
in the main sequence,
running from upper left to
lower right, with numbers
increasing toward the faint,
red end. Above the main
sequence are many giants
and rare supergiants. Tiny,
faint white dwarfs lie
across the bottom.

Red Supergiant
Betelgeuse in Orion is a red
supergiant, some 15,000
times as luminous as the
Sun and with a greater
diameter than Earth's orbit.

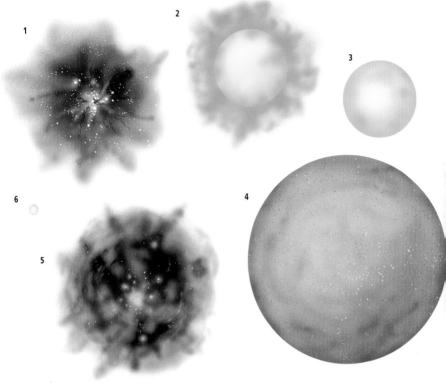

Life of a Star

Stars begin life as a large cloud of gas and dust (1) and progress through an intermediate stage (2) to become a fully fledged star, burning hydrogen for fuel (3). When it runs out of fuel, an average-size star swells up and cools to become a red giant (4). Eventually, the giant's outer layers are driven off to create a planetary nebula (5). The nebula dissipates and the core of the star is left behind as a cool, fading white dwarf (6).

numbered 0 to 9. The Sun, which is a G2 main-sequence star, lies roughly in the middle of the main-sequence region of the diagram and is referred to as a yellow dwarf.

A STAR'S FATE

A star's birth and lifespan depend greatly on its mass. Massive stars are in a hurry. They take the shortest time to form—on the order of hundreds of thousands of years compared to the Sun's tens of millions of years—but have relatively brief lives. A star that is eight times more massive than the Sun will shine for only 40 million years, while a star like the Sun can live for 10 or 11 billion years.

Death, too, depends on mass. After it expends its nuclear fuel, a massive star goes out in a blaze of glory, as a supernova (see p. 176); the death of a smaller star, such as the Sun, is much more subdued.

The Sun entered the main sequence 4.5 billion years ago and has been shining steadily ever since, fusing hydrogen atoms in its core into helium. As the helium builds up, the Sun's luminosity will increase, probably within the next billion to two billion years. To put it mildly, this is not good news for planet Earth, where the increased heat will produce a Venus-like runaway greenhouse effect, evaporating the oceans and increasing clouds.

As internal temperatures rise, the Sun's gaseous envelope will begin to expand. As it does so, our star's surface will cool and grow redder. In about 6.5 billion years, the Sun will become a red giant. At some point thereafter, the amount of helium in the core will reach a critical threshold and ignite in a "helium flash." This will cause the Sun to shrink slightly, then balloon outward even further. Eventually, its outer atmosphere will drift out into space, forming a planetary nebula. Left behind is a white dwarf, an

Helix Nebula
This image of the Helix Nebula shows the final days of a star like our Sun. The outer parts of the star expanded to create a glowing cloud—a planetary nebula. The intersecting rings of the nebula span about 1.5 light-years.

VARIABLE STARS

The majority of stars shine steadily, their brightness hardly fluctuating. A star like this is in perfect balance between gravity, which acts to collapse the star, and internal heat, which acts to make it expand. Should one force outbalance the other, the star would become unstable and vary in brightness.

Variable stars pulsate between a period of maximum and minimum brightness, some regularly, some irregularly. When the star pulsates outward to its maximum, it expands beyond the point where gravity and gas pressure are in equilibrium. Gravity then slows and reverses the expansion. As the star contracts, it goes to the other extreme,

shrinking below the equilibrium point until the increase in internal gas pressure slows and counteracts the contraction.

Some pulsating variables act as celestial yardsticks, allowing accurate measurement of vast distances. Cepheid and RR Lyrae stars vary in brightness with great regularity. A typical "light curve" (the red line in the diagram) shows a Cepheid's brightness peaking quickly and then declining over a period, typically, of a few days. Astronomers know that these stars show a direct relationship between the period of variability and their absolute magnitude: the longer the star's period, the more intrinsically luminous it is. The star's distance from Earth is calculated by comparing absolute to apparent magnitude.

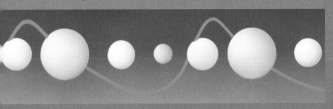

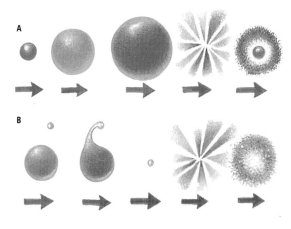

A

B

and indeed, may repeat the behavior thousands of years later. A supernova is more final. It is the catastrophic death of a star.

Most stars have at least one companion star. Novas occur in a binary star system in which one star is a white dwarf and the other is an ordinary star. The white dwarf has greater gravity than its companion, and is able to siphon hydrogen-rich gas from it. The gas swirls into a disk around the white dwarf before spiraling down onto the dwarf's surface. When enough gas has accumulated on the surface—after 10,000 to 100,000 years—the white dwarf erupts in a thermonuclear explosion, causing the spectacular nova outburst.

A supernova is one of the most violent explosions in the universe. It is a hundred times more luminous than a nova and can outshine hundreds of billions of stars in its galaxy. These stellar fireballs are instrumental in distributing heavy elements, such as iron and nickel, throughout the galaxy.

Two types of supernova take place, depending on what kind of star is involved. The first occurs in a binary system when gas from

Going Supernova
A star at least eight times more massive than the Sun starts as a blue-white star (A). As it runs out of fuel, it swells and cools. When its fuel is spent, its core collapses and it explodes in a supernova, leaving behind a neutron star or black hole (see p. 185) and an expanding supernova remnant of gaseous debris. In a binary system (B), a white dwarf sometimes drags material from its companion star, becomes unstable, and explodes, destroying the white dwarf.

intensely hot core no larger than Earth. The Sun's exhaled gas will drift into space where, millions of years later, it will be recycled into a new generation of stars.

VIOLENT OUTBURSTS: NOVAS AND SUPERNOVAS

Novas and supernovas are superficially similar: both involve stars that show unpredictable and significant increases in brightness. In truth, however, they are very different stellar beasts. A nova is a star that suddenly increases in brightness by as much as 10 magnitudes then declines to a value close to its original magnitude. The star is not destroyed

Nova in Cygnus
Nova Cygni 1992 erupted in February 1992. A Hubble image taken 467 days later (left) shows a ring of gas around the star. Seven months later (right), the ring had expanded.

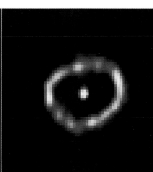

A SUPERNOVA REMNANT IN CASSIOPEIA

When a star in the constellation Cassiopeia exploded as a supernova nearly 10,000 years ago, a huge shell of gas raced out into space. Today, this still-expanding supernova remnant—Cassiopeia A—is too faint to see with amateur equipment, but its energy can be detected by radio telescopes.

Radio images of the cloud, the brightest radio-wavelength source outside the Solar System, show gas racing away from the spot where the star exploded. By calculating this speed and the distance traveled by the gas since the explosion, astronomers estimate that the light from the explosion reached Earth around the year 1680, creating a magnitude 5 star.

Supernova 1987A

Naked-eye supernovas are very rare. The one that erupted in early 1987 in the Large Magellanic Cloud occurred 400 years after the previous one. These before and after photos show how SN 1987A started out as just another faint star (left) to become something quite out of the ordinary (right). The star actually erupted some 165,000 years ago, the light having taken that long to reach us.

a moderately massive star accumulates onto a white dwarf, pushing the white dwarf past the point where its internal pressure can counteract gravity. The infalling gas triggers a thermonuclear explosion that probably disrupts the entire star system.

If a star is particularly massive—about eight times as massive as the Sun—and its nuclear fuel has been expended, it cannot support its own outer layers. The core first collapses then rebounds catastrophically, blowing off the star's outer layers.

Eta Carinae: the Next Supernova?

Eight thousand light-years away from us is the Eta Carinae Nebula, a complex of young, hot stars. Between 1835 and 1845, observers on Earth watched as a star in its midst, Eta Carinae, grew in brightness to outshine every star in the night sky except Sirius. It is now very faint and cannot be seen without binoculars. Eta Carinae may be a very young star experiencing a series of outbursts. On the other hand, it may be a very old one that is about to become a supernova, in which case it will be one of the brightest stellar deaths ever seen in our Galaxy.

Cosmic Explosion

The explosion that caused Eta Carinae's surge in brightness in the 1800s created these two huge, billowing clouds of gas and dust, seen in a Hubble Space Telescope image. Astronomers estimate that the clouds are racing away from the star—the bright white spot at the center—at about 1.5 million miles per hour (2.4 million km/h). The amount of gas and dust ejected is hard to grasp—it is estimated to be more than 500 times the diameter of our Solar System.

Shape of Things to Come?

Perhaps Eta Carinae will look something like this in the far-distant future. The Cygnus Loop (also called the Veil Nebula) lies 2,500 light-years away. It is the expanding lacy blastwave from a supernova that exploded about 15,000 years ago.

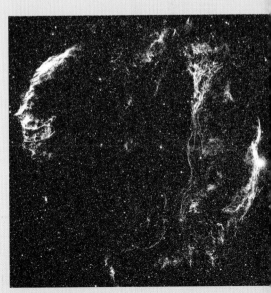

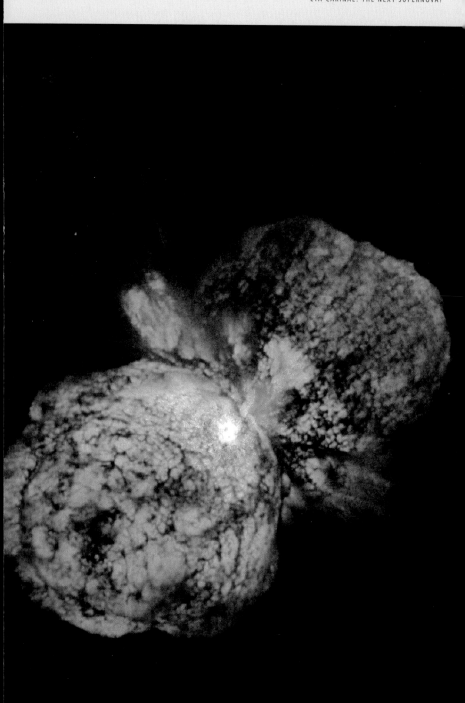

Double and Multiple Stars

Our Sun shines by itself in proud isolation, many light-years from the nearest other star. But the Sun is in the minority. Most stars are not alone, but occur as pairs, triples, or even clusters.

Alnitak's Companion
Alnitak, one of the stars in Orion's belt, almost overshadows its close and much dimmer companion.

Double-star System
The members of a double-star system orbit around a common center of gravity. If one has more mass than the other, the center of gravity is nearest the more massive star.

Astronomers calculate that more than half of all stars in the night sky harbor one or more companion stars. These star systems are thought to form when several different parts of the parent cloud of gas and dust collapse at once.

STELLAR COMPANIONS

Double stars are two stars that appear near each other in space. If the stars are gravitationally bound to each other, they are called physical doubles or visual

SPECTROSCOPIC BINARIES

If the stars in a binary system are very close to each other, their orbital motions create variations in their line-of-sight velocity, like dancers of different weight pivoting around their center of gravity. Astronomers use a spectroscope to detect these shifts, from which the relative velocities and masses of the two stars can be calculated. This kind of system is called a spectroscopic binary.

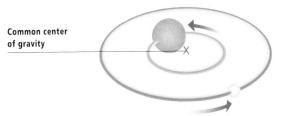

Common center of gravity

binaries, and they move together through space and slowly orbit one another. Mizar, the star marking the bend in the handle of the Big Dipper, is a physical double. Of the two stars, the

Capella

Capella, the brightest star in the constellation Auriga, is a pair of yellow giants, each with a companion star. The stars are too close to separate in a conventional telescope, but the giants' motion, shown here, can be seen with a special imaging technique. The images were taken two weeks apart.

brighter member is referred to as the primary while the fainter one is called the companion or the secondary.

Stars with two or more companions are known as multiple stars. Easily visible in a small telescope are the four prominent young stars that make up the system called the Trapezium, in the center of the Orion Nebula. The nearest bright star to Earth, Alpha Centauri, is a triple-star system.

Sometimes, a chance alignment makes it seem that two stars are close to each other, when in fact they are widely separated in space. The star Deneb in the constellation Cygnus is one such example. These optical doubles are not bound to one another by gravity.

ECLIPSING BINARIES

A binary system may be aligned in such a way that, as seen from Earth, one star passes in front of the other, and then behind it. This causes a periodic variation in brightness. Astronomers use these eclipsing binaries as gravitational laboratories, monitoring the variations of light in the star

system to find out the mass, diameter, and orbital motion of the stars.

One of the most famous eclipsing binaries is Algol in the constellation Perseus. The two stars revolve about each other every 2.8 days. A larger, fainter star cuts off the light of the smaller, brighter member, causing the magnitude to fall from 2.1 to 3.3. The changes are readily visible to the naked eye.

Eclipsing Binary

In eclipsing binaries such as Algol, two stars orbit each other. If the orbit is angled just right, when one star passes behind the other, its light is blocked and the total amount of light from the pair dims.

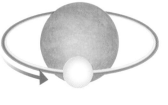

Bright white star blocks some light from orange star— medium brightness

White star beside orange star— maximum brightness

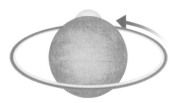

Orange star blocks all light from white star—minimum brightness

Stellar Nurseries and Graveyards

Nebulas are gaseous "clouds" scattered between the stars. Some of these often beautiful objects harbor the remains of dead stars; others are places where stars are being born.

Emission Nebula
Emission nebulas are among the most beautiful of all deep-sky objects. This is the Orion Nebula, which is lit up by four bright central stars known as the Trapezium cluster.

Reflection Nebula
Starlight reflecting from tiny dust grains produces this reflection nebula, seen in the constellation Corona Australis.

Astronomers originally used the word "nebula" to describe anything that appeared blurry or cloudlike through a telescope. Many of these clouds, it turned out, had distinct spiral or oval shapes and were later recognized to be galaxies. But a large number had no particular shape and appeared to have stars embedded within them. These complexes of interstellar gas and dust can be observed as either a luminous patch on the sky or a dark cloud against the background stars.

TYPES OF NEBULAS

There are three main types of nebulas: emission, reflection, and dark. Emission nebulas (also known as HII, pronounced "H-two," regions) are those whose atoms are excited, or ionized, by ultraviolet radiation emitted by hot stars from within. This process makes nebulas glow brightly, and shine colorfully in photos. One of the brightest is the Orion Nebula, located in the sword of stars hanging from the belt of

Orion the Hunter. Emission nebulas are associated with both star formation and star death.

Reflection nebulas do not shine on their own power but y the light of adjacent stars scattering off dust grains in the cloud. The tiny dust grains reflect blue light more effectively than red light, so these nebulas often appear blue in color. Sometimes, both emission and reflection regions exist in the same nebula —the Trifid Nebula in Sagittarius a good example.

Dark nebulas are clouds of gas and dust dense enough to block the light from stars in the background—they show up as amor-phous, dark silhouettes (such as the Horsehead Nebula in Orion) or as "holes" in space (such as the Coalsack in Crux). Many dark nebulas obscure regions of newly formed stars.

PLANETARY NEBULAS

Quite different to the other clouds of gas and dust are the planetary nebulas. Through a telescope, these nebulas often appear as small disks—not unlike a planet, hence the name. These expanding rings or shells of gas have nothing to do with planets; nor are they associated with star-birth. They are the final stages in the life of small-mass stars.

Dark Nebula
The Horsehead Nebula is a dense black cloud of dust that can be seen only because it blots out some of the light coming from the glowing streamers of an emission nebula.

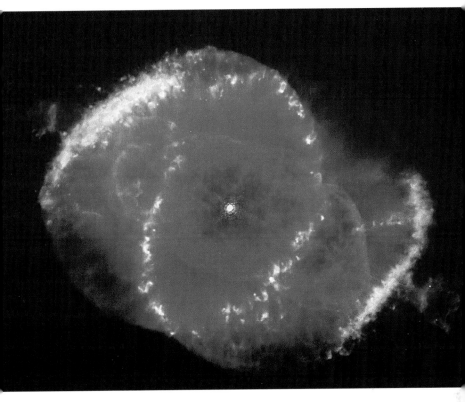

Planetary Nebula
The Cat's-Eye Nebula is the wreckage of a star that threw off its outer layers about a thousand years ago. Its unusual formation could be due to the gravitational effects of a companion star which survived the eruption.

A smallish star—one not more than eight times as massive as the Sun—starts to die when its store of nuclear fuel at the core runs out. The core contracts and grows hotter, and the increased radiation causes the star's outer layers to balloon. Gravity's hold on them is weak, and the stellar "wind" puffs the layers into space.

Something else besides gas remains behind. The dense hot core of the star is now a white dwarf, which emits radiation into the surrounding shell, causing it to glow. The Ring Nebula in the constellation Lyra and the Dumbbell Nebula in Vulpecula are two of the finest planetary nebulas in the sky.

SUPERNOVA REMNANTS

Like planetary nebulas, supernova remnants are expanding shells of gas—but they are produced by massive stars that have exploded in a catastrophic eruption (see p. 176). Their remains are usually fainter and less symmetrical than planetary nebulas.

Supernova remnants less than a thousand years old are strong sources of radio waves and X rays. A prominent example is the Crab Nebula in Taurus, first observed as a "guest star" in July or August of AD 1054 by Chinese and Japanese astronomers. This young remnant is also illuminated by a central pulsar (see box, p. 185) created in the explosion.

Supernova Remnant
The Crab Nebula is the gaseous remains of the explosion of a star witnessed in AD 1054. As a young supernova remnant, it has a more regular, ringlike shape than older examples. At the Crab's heart is a pulsar, the fastest and most energetic star formed from a supernova.

As supernova remnants age, they become less regular in shape, appearing as huge rings, filaments, or arcs of tenuous gas; they are also associated with radio and X-ray emissions. Prominent in the constellation Vela are the lacy remains of a massive star that is thought to have exploded about 12,000 years ago.

NEUTRON STARS, PULSARS, AND BLACK HOLES

A supernova is such a huge explosion that it is hard to believe that anything could be left behind apart from drifting gaseous material. In some cases, however, the burnt-out core of a supernova survives but is so massive—at least 1.4 times the mass of the Sun—that it collapses into a neutron star. This object is incredibly compact and very dense, with a diameter of perhaps 10 miles (15 km) but having a mass equal to the Sun's.

A neutron star spins rapidly and emits radio waves. If the waves sweep across Earth, they may be detected as cyclical "pulses," resembling the flashes produced by a lighthouse. This type of object is a pulsar. The pulsar in the Crab Nebula (seen in the X-ray image below as the bright spot left of center) flashes on and off 30 times a second.

What happens to a truly enormous star at its death? If the core of a star has a mass three or more times that of the Sun, it can collapse into a black hole—an object so dense that not even light can escape its gravity. To "see" a black hole, astronomers turn their attention to companion stars. If a black hole and a normal companion are close enough together, gas is siphoned from the normal star into a hot accretion disk surrounding the black hole. X rays emitted during this process can be detected by astronomers. Also, the normal star orbits around the system's center of gravity at a rate determined by the combined mass of both bodies—another telltale signature of a black hole.

The first possible black hole to be identified was Cygnus X-1. It consists of a visible star with a mass 20 times the Sun's orbiting with an invisible object half its mass.

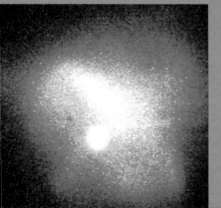

A Look Inside
the Eagle Nebula

Stars are forming inside the Eagle Nebula in the constellation Serpens, 7,000 light-years from Earth. In April 1995, the Hubble Space Telescope recorded a stunning image of this nebula, revealing pillar-like structures of cool hydrogen gas and billowy dust festooned with young stars. The pillars, which protrude about 1 light-year from an especially dense cloud of molecular hydrogen gas, are being worn away by an unceasing blast of ultraviolet radiation from hot, young stars. This exposes denser, dusty globules of gas, where stars are being formed, and lights up the surfaces of both the pillars and the streamers of gas driven away from them.

Pillars of Dust and Gas

Pillars of dust and gas are clearly seen in this spectacular Hubble image of the central part of the Eagle Nebula. The light from newborn stars is eroding the gas in the pillars, much as wind moves dust on Earth. Just visible at the tip of the pillar on the left are small globules of dusty gas—each about the size of the Solar System—in which stars are forming.

Star-lit Eagle

This photo of the Eagle Nebula was taken by an Earth-based telescope and shows dust clouds and gas being lit up by the light from young stars. The Hubble Space Telescope peered into a small region at the center to produce the image on the facing page.

Other Solar Systems

What was once pure speculation is now unshakable certainty: planets orbit stars other than the Sun. In fact, more planets are known to exist outside our Solar System than inside it, and more are being discovered all the time.

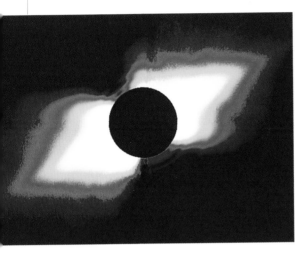

Hidden Planets?
The disk around the star Beta Pictoris is slightly warped, possibly because of the gravitational pull exerted by a planet or two hidden close to the star.

Wobbling Star
Close observation may reveal a slight wobble in a star's path—a telltale sign of the presence of a planet.

Star
Planet

The first extrasolar planet was discovered in October 1995 by two Swiss astronomers, Michel Mayor and Didier Queloz. They found a planet about half the size of Jupiter orbiting a star called 51 Pegasi in the constellation Pegasus. Since then, more than a hundred planets have been discovered, and there are many more tentative findings.

HOW TO FIND
AN EXTRASOLAR PLANET

You cannot see a planet outside our Solar System. They reflect too little light, and besides, they lie within the glare of their host star—like a gnat beside a spotlight. The only way to detect their presence is by indirect means. Astronomers have two successful methods, both of which involve closely examining the star for the tiny effects of an orbiting planet.

As a planet orbits, its gravity causes a minute movement in the parent star. As seen from Earth, the star "wobbles," moving backward and forward. To detect this motion, the most commonly used technique tries to discover changes in the color of a star's light. The color shifts toward the blue end of the spectrum as the star moves toward Earth; when it recedes it shifts toward the red. The pattern of color changes can tell astronomers the planet's orbital period and, given a few assumptions, its mass.

The other technique involves taking precise measurements over time of a star's position in relation to the more distant background stars. Changes in the star's position can indicate the presence of a planet. They can also reveal the planet's orbital period and, again using some assumptions, its mass.

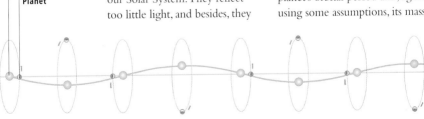

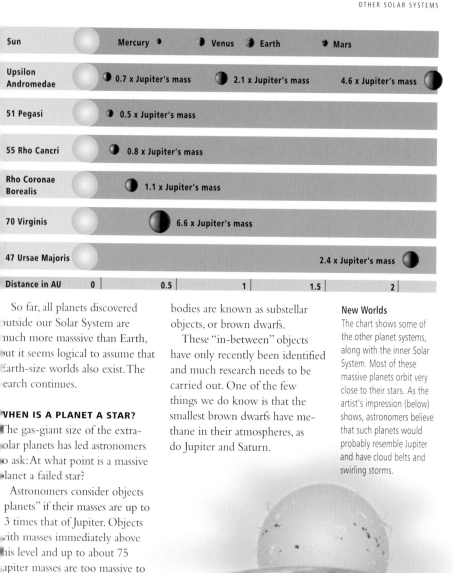

Sun		Mercury	Venus	Earth	Mars
Upsilon Andromedae		0.7 x Jupiter's mass	2.1 x Jupiter's mass	4.6 x Jupiter's mass	
51 Pegasi		0.5 x Jupiter's mass			
55 Rho Cancri		0.8 x Jupiter's mass			
Rho Coronae Borealis		1.1 x Jupiter's mass			
70 Virginis		6.6 x Jupiter's mass			
47 Ursae Majoris				2.4 x Jupiter's mass	
Distance in AU	0	0.5	1	1.5	2

So far, all planets discovered outside our Solar System are much more masssive than Earth, but it seems logical to assume that Earth-size worlds also exist. The search continues.

WHEN IS A PLANET A STAR?

The gas-giant size of the extra-solar planets has led astronomers to ask: At what point is a massive planet a failed star?

Astronomers consider objects "planets" if their masses are up to 13 times that of Jupiter. Objects with masses immediately above this level and up to about 75 Jupiter masses are too massive to be planets, but they are not stars either—they cannot sustain hydrogen fusion. These bodies are known as substellar objects, or brown dwarfs.

These "in-between" objects have only recently been identified and much research needs to be carried out. One of the few things we do know is that the smallest brown dwarfs have methane in their atmospheres, as do Jupiter and Saturn.

New Worlds

The chart shows some of the other planet systems, along with the inner Solar System. Most of these massive planets orbit very close to their stars. As the artist's impression (below) shows, astronomers believe that such planets would probably resemble Jupiter and have cloud belts and swirling storms.

Cities of Stars

Most stars are clustered in groups that may be hundreds of thousands strong. Star clusters give professional astronomers insights into the lives of stars, and provide backyard observers with some of the most stunning sights in the sky.

Pleiades
Though only six or seven stars can be seen by the naked eye, the Pleiades star cluster actually contains some 3,000 stars. This open cluster lies about 400 light-years away in the constellation Taurus.

Aging Open Cluster
When an open cluster forms, it is less than about 30 light-years in diameter and has a broad range of stars. In time, the members drift apart (below) and evolve as all stars do. Eventually, the cluster becomes hard to tell apart from the rest of the galaxy's stars (bottom).

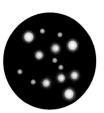

Star systems not only come in pairs and multiples, but also in clusters that range in size from tens of stars to more than a million. Mutual gravity holds them together—a bond that may be broken only by an outside force—and they drift through space in the same direction and speed, much like a school of fish.

The stars in a cluster are of roughly the same age and composition, having formed from the same interstellar cloud. Clusters vary in age from a few million to billions of years.

OPEN CLUSTERS
Loose assemblages of stars, containing at most a few thousand members, are called open or galactic clusters. Some 1,200 open clusters are known in the Milky Way Galaxy, and many have been observed in nearby galaxies. The smallest are just a few light-years in diameter; the biggest may be a hundred times larger. Examples include the Hyades in Taurus, the Double Cluster in Perseus, and the Praesepe or Beehive Cluster in Cancer.

Open clusters are found along the spiral arms of the thin galactic plane, which is the main location for star formation. Most of these stars are younger than our Sun, some being among the youngest stars we can see.

In cosmic terms, open clusters have short lives. Few are thought to survive the gravitational

TELLING A CLUSTER'S AGE

A star cluster's age can be determined by studying the color of the light emitted by the cluster's stars. Older stars have a distinctive yellow to reddish color and shine at a particular brightness. Using stellar evolution models that show how old a cluster must be with a given color and luminosity, astronomers can narrow down its age to within a few million years.

erturbations of other stars and nterstellar clouds for more than few hundred million years.

LOBULAR CLUSTERS

n a small telescope, a globular luster looks like a fuzzy ball. t takes a larger instrument to esolve it into what it truly is—a reat conglomeration of untold thousands of stars. Some globulars, in fact, have more than a million members. These stars are packed tightly together, in an area only 30 to 100 light-years across.

Most of the globulars we see are located in a halo surrounding our Galaxy, and similar halos of globulars have been observed around other galaxies. The most spectacular cluster of all is Omega Centauri, with its million or so stars of 11th magnitude or greater. It is seen best from the Southern Hemisphere. The brightest globular in the northern sky is the Hercules Cluster, also known as M 13.

Some of the oldest stars in our Galaxy belong to globular clusters. Observations in 1997 by the Hipparcos satellite revealed that the average age of most globular clusters is about 11.5 billion years—nearly the age of the universe itself.

Aging Globular Cluster
Because they have so much mass and gravity, globular clusters hold onto most of their stars for a long time. A young globular (top) has many white-hot stars, plus yellow stars and dim red dwarfs. In an old globular (above), the stars in the original population have evolved, becoming red giants and white dwarfs.

Omega Centauri
Some 17,000 light years away, Omega Centauri is among the nearest globular clusters to Earth. Its myriad stars are packed so tightly that near the center, they are typically only one-tenth of a light-year apart. The cluster can be seen by the unaided eye as a 4th-magnitude spot of light.

The Milky Way

From Earth, the Milky Way appears like a gossamer band of light or a high, thin cloud, for which it is often mistaken. In fact, it is a galaxy—our Galaxy—seen from the inside.

The Solar System is embedded within a system of stars known as the Milky Way, or the Galaxy. If we could transport ourselves outside our Galaxy, we would see a rotating disk some 100,000 light-years in diameter containing hundreds of billions of stars with lanes of dark dust and glowing gas winding out of a central hub like a pinwheel. The Sun would appear as an insignificant yellow star near the mid-plane of the disk some 27,000 light-years from the center and we would have to wait 230 million years for it to complete one galactic orbit.

Inside View

Away from city lights, the Milky Way is a spectacular sight. This star-strewn band is our view of the galactic disk, the thinnest part of our Galaxy. The photo is taken looking toward the center of the Galaxy, a place we can never see directly because of obscuring clouds of dust.

A BLUEPRINT OF THE MILKY WAY

The Milky Way is a spiral galaxy, a class that comprises the largest and most symmetrically shaped structures in the universe. Young, hot stars shine within open clusters along the spiral arms in the galactic plane; some are just emerging from their interstellar shells. Encompassing the center of our spiral is the galactic bulge, made up mainly of old red stars. The entire system is enveloped in a halo made up of old stars and dotted with globular clusters.

Astronomers are still unclear about precisely how many spiral

ns the Milky Way has, but some thorities map as many as five: e Orion arm—which contains e Sun and the bright stars we e in the sky—the Perseus, the gittarius, the Crux-Centaurus, d the Cygnus. Both the Perseus d Sagittarius arms lie at distances of about 6,500 light-years.

ALACTIC MYSTERIES

ne Galaxy is our own backyard, d we know much about it. t it harbors two intriguing ysteries, one at its center, the her on its outermost fringes. We cannot see the center of ur Galaxy—dust obscures the ew—but infrared and radio aves are able to penetrate the ze. Observations at these wavengths have allowed astronomers measure the velocities of stars biting within 5 light-years of e center. Based on these veloces, astronomers calculate that object with a mass about 2.6 illion times that of the Sun lies

at the center. They believe it to be a supermassive black hole.

Surrounding the Milky Way is a dark outer halo of unknown composition containing 10 times the mass of all the known stars in the Galaxy. This halo is invisible, yet we know it exists by the gravitational drag it exerts on distant globular clusters in the inner halo and on nearby dwarf galaxies. No one knows for sure, but astronomers theorize that the dark matter consists of compact objects, such as white dwarfs or neutron stars, or a "soup" of elementary particles, or perhaps a combination of the two.

Outside Views
If we could see our Galaxy from the outside, it might look like these two views. Face-on, note its spiral structure (left) and edge-on, see how thin the spiral disk is (right).

The Glow of Home
This infrared image highlights the warmest areas of the Milky Way. The infrared radiation comes mostly from the dust that lies throughout the disk. The dust glows with the heat of starlight, and clumps of dust contain new stars.

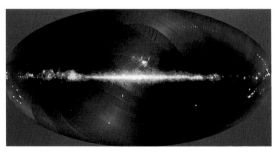

Galaxies

Galaxies are immense and remote systems of interstellar gas and dust containing billions of stars. There are thought to be billions of them, in almost limitless variety.

Spiral Galaxy
NGC 2997 in Antlia is an impressive sight, with its arms traced out by blue stars, pink hydrogen clouds, and dust. This spiral galaxy is smaller than our own but still weighs about 100 billion times as much as the Sun. It lies about 55 million light-years away.

Galaxies come in a range of shapes, from pinwheels and spheres to ellipsoids and shapeless blobs. Astronomers classify them by shape, to better understand galactic evolution and structure.

The galaxy classification system used by many astronomers today originated in 1926 when Edwin Hubble sorted galaxies into three broad categories: ellipticals, spirals, and barred spirals. Two other categories, lenticulars and irregulars, were added later.

ELLIPTICAL GALAXIES
Elliptical galaxies tend to be shaped like spheres, although many appear flattened or lens-shaped. The largest and rarest ellipticals have diameters of 100,000 light-years or more and have a mass 100 trillion times that of the Sun. Of the thousand brightest galaxies in the sky, large ellipticals—such as M 86 and M 87 in Virgo—make up about 20 percent. More common are the dwarf ellipticals, which may

Elliptical Galaxy
Elliptical galaxies vary in size from dwarfs to giants. M 87 in Virgo is, without doubt, a giant—it is thought to have a mass equivalent to more than one trillion Suns.

have a diameter of only 1,000 light-years or so and have a mass few million times the Sun's.

Scientists think that elliptical galaxies form in dense regions and reach their mature shape very quickly following a vigorous burst of star formation. They then retire early and for the next 10 to 11 billion years quietly live out their lives, burning steadily away like gas street lamps in a London fog.

SPIRAL AND BARRED SPIRAL GALAXIES

Most bright galaxies, about 75 percent, are spirals. These range in size from 15,000 to 150,000 light-years across and may contain anywhere between 10 billion and 10 trillion solar masses. The spiral arms unwind from a bright central region, called the nucleus, and wrap around the disk. Young, hot stars trace out the spiral arm structure like lights around a Christmas tree. Dark lanes of interstellar dust, bright nebulas, and star clusters are distributed throughout the galaxy's thin disk. A spiral galaxy's hub is called the nuclear bulge, a region of predominantly old red stars. The

Barred Spiral Galaxy
In barred spirals, the arms emerge from a bright bar that sits across the center of the galaxy. The beautiful NGC 2442 in Volans presents a classic barred structure. It seems that many ordinary spirals also have indistinct bars.

bulges are prominent in some spirals and almost nonexistent in others. Our Galaxy and the neighboring Andromeda Galaxy are typical spirals.

Astronomers theorize that spiral galaxies probably took much longer to evolve than ellipticals. Most were formed from gas fragments or collapsed from disks in turbulent regions that induced rotation in the still-forming galaxy. Judging from their great range of different shapes, many have been torn apart and reshaped by multiple mergers and sideswipes with other galaxies.

Barred spirals are much rarer than true spirals. In this type, the bright stars and hot gas of the inner regions extend for thousands of light-years from either side of the center in a straight "bar" before wrapping back around the galaxy in the form of arms. In obvious cases, the general appearance of each arm is something like a scimitar or scythe blade.

EVOLVING GALAXIES
Astronomers once believed that one type of galaxy evolved into another over time. An elliptical galaxy, for example, gradually became a spiral. The current view is that galaxies retain the same shape throughout their lives—barring, that is, collisions with other galaxies or some other catastrophic event.

195

Galaxies in Collision

Sometimes two or more galaxies get too close to each other and collide or merge. Computer models of galaxy collisions and near misses produce the same distinctive rings, "tails," and arcs of gas, dust, and stars that are evident in some galaxy groups. The theory goes that two or more spiral galaxies attracted by their gravity can get trapped in tight orbits around each other and collide. A second and third collision may follow before the galaxies consolidate into a single disorganized accumulation of stars. The whole process takes a few hundred million years. Some astronomers believe that the end result of a merger between two disk galaxies is an elliptical galaxy.

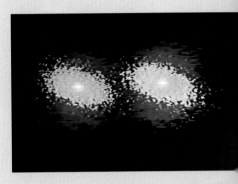

Computer Mode
Above: Computer modeling allow astronomers to test theories again observations. This simulation shows tw spiral galaxies about to collide. An elliptic galaxy may be the eventual outcom

Collision-cours
Facing page, above: When galaxies collid they pass right through each other becau their stars lie far apart. But the gravitation pull of such a collision usually triggers ea galaxy's gas clouds to begin making ne stars. The Antennae galaxies, which bega colliding 500 million years ago, featu bright areas of young sta

Ring of Sta
Facing page, below: The Cartwheel Galaxy the spectacular aftermath of an ancie collision. A small galaxy, perhaps one of tho on the right, cannoned into the larger or sending a wave of energy out into space li a ripple on a pond. This created several billi new stars in an expanding ring that is nc large enough to contain our entire Gala

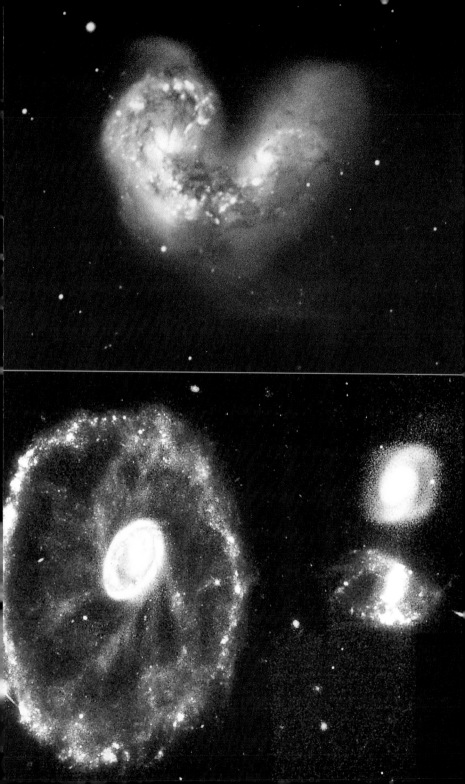

Galaxy Clusters and Superclusters

Galaxies occur in clusters and also in clusters of clusters of incredible size. Studying how such enormous structures came into being is helping astronomers understand how the universe formed and evolved.

Galaxies are sometimes referred to as "island universes," which implies that they are self-contained objects surrounded by immense expanses of space. They *are* self-contained—by the gravitational attraction of their components—but they do not exist in total isolation. Most galaxies, including our own, are part of larger groups called clusters of galaxies.

CLUSTERS RICH AND POOR

Some galaxies belong to relatively small clusters only a few hundred thousand light-years across; others amass in thousand-strong conglomerations that may extend for tens of millions of light-years. These immense groupings are bound together by gravity and move through space as one.

Over the past several decades, galaxy surveys have turned up nearly 3,000 clusters within 4 billion light-years. Two distinct classes emerge: rich and poor. Rich clusters contain thousands of galaxies distributed over roughly 10 million light-years in diameter. The nearest moderately rich cluster of galaxies lies in the constellation Virgo. This mighty assemblage has more than 2,000 member galaxies located about 50 million light-years away and covers an area of sky some 120 square degrees.

Other examples of rich clusters include the Coma cluster, 230 million light-years away, with about a thousand large galaxies and 20,000 or 30,000 smaller systems, and the Hercules cluster, between 300 and 700 million light-years distant and containing tens of thousands of galaxies.

Our galaxy belongs to a poor cluster known as the Local Group. The Milky Way Galaxy and the Andromeda Galaxy are

Local Group

The Local Group is the cluster of galaxies to which our Galaxy belongs. The 17 largest members are illustrated here, with the Milky Way, Andromeda, and Pinwheel galaxies the largest of all. Most of the small galaxies orbit around either the Milky Way or Andromeda in the same way that planets orbit a star, and the rest are affected by the giants' huge gravitational pull.

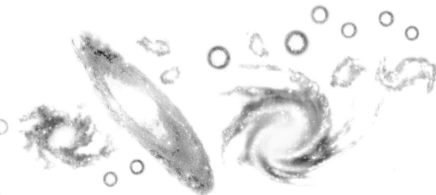

COUNTING THE LOCAL GROUP
The Local Group has at least 35 member galaxies, but we may never know the exact number because dust in our Galaxy obscures part of the view. Any remaining members would be small and dim.

Rich clusters usually have a hierarchical structure, with a center containing almost exclusively giant ellipticals and an outer region of gas-poor disk galaxies. It is thought that collisions and mergers probably stripped gas from other galaxies in the cluster and that the gas sank toward the cluster's center, increasing the mass of the giant ellipticals there. Rich clusters also tend to be symmetrical and spherical in appearance, while poor ones are amorphous, possessing less gravity to give a regular shape.

SUPERCLUSTERS
Clusters of galaxies themselves belong to larger groupings known

Satellite Galaxies
Two medium-size members of the Local Group can be seen together in southern skies. The Small (left) and Large (right) Magellanic Clouds look like two detached parts of the Milky Way. They are in fact two irregular satellite galaxies of our own Milky Way.

he largest members of this humble grouping, which measures about 8 million light-years across and contains at least 43 galaxies, including the Large and Small Magellanic Clouds and the Pinwheel Galaxy. The majority of the group's members are dwarf elliptical and irregular galaxies containing only about a million stars each..

Virgo Cluster
Two elliptical galaxies,
M 86 (center) and M 84
(right), take pride of place
in this view of a small part
of the Virgo cluster. At
about 50 million light-years
away, it is the closest
collection of galaxies to us.
Some 2,000 members
strong and packing a huge
gravitational punch, it
dominates our neighbor-
hood, lying at the center
of the Local Supercluster.

as superclusters. Working in the
late 1950s and early 1960s, the
U.S. astronomer George Abell
found 50 superclusters, each
containing, on average, about a
dozen rich clusters. The largest is
truly colossal, with a diameter on
the order of 300 million light-
years. By contrast, the supercluster
to which our Local Group
belongs is a trifling 100 million
light-years across, and contains
one rich cluster—the Virgo
cluster—and several poor ones.

Which came first—clusters or
superclusters? Astronomers are
divided. The proponents of the
"top-down" theory argue that
superclusters of galaxies formed
first, then gradually fragmented
into smaller clusters. Champions
of the "bottom-up" theory con-
tend that galaxies accumulated
from small conglomerates of
matter and slowly grew into
clusters and finally superclusters.
Astronomers are at the edge of
their knowledge here, and much

Fornax Cluster
Most of the galaxies in the
Fornax cluster, 60 million
light-years away, are
ellipticals, although the
largest member is the
barred spiral at top right of
the photo, NGC 1365.

esearch needs to be done before
hey can give a lasting explanation
f these structures.

HE FACE OF THE UNIVERSE

ver since George Abell's time,
stronomers have been busy map-
ing the superclusters to create a
ast cosmic canvas, and pondering
he meanings of the large-scale
attern they see. This pattern is
nything but uniform: super-
lusters are arranged in vast chains
hat meander around enormous
racts of apparently empty space.

What brought these structures
nto existence? And when did
hey form? Did galaxies come
nto being a billion years after the
niverse began at the Big Bang
r several billions of years later?
uch questions touch on funda-
nental assumptions about the

creation of the universe, and as
yet we have no definitive answers,
although there are plenty of
theories (see p. 212).

New, more ambitious surveys
may provide much-needed
insights into large-scale structure.
One of these, the U.S.–Japanese
Sloan Digital Sky Survey, is now
underway using a purpose-built
telescope in New Mexico. It will
provide the information needed
to plot the positions and give the
absolute brightnesses of more
than 100 million celestial objects,
including a million galaxies and
100,000 quasars. Such observations will allow astronomers to pit
their theories about large-scale
structure against real data, and
perhaps lead to an understanding
of how the universe attained its
complex shape.

Galaxy Surveyor
The Sloan Digital Sky
Survey aims to map one
quarter of the night sky. To
take in as much of the sky
as possible at a time, the
survey uses a specially
designed telescope which
incorporates two instru-
ments: a spectroscope and
what may be the most
complex camera ever built.

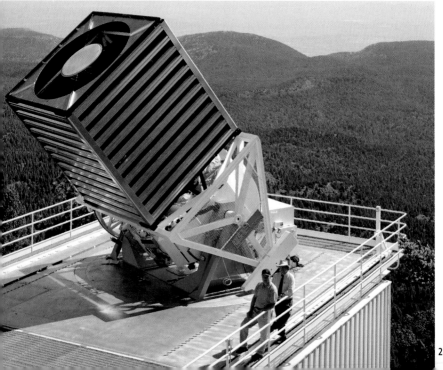

The Big
Questions

How did the universe form? When will it end? Does life exist outside Earth? The "big questions" about the universe are the hardest to answer, but recent years have produced remarkable observations to match some amazing theories.

"Now, my own suspicion is that the universe is not only queerer than we suppose, but queerer than we can suppose."

Possible Worlds, J. B. S. Haldane
Scottish mathematical biologist, 1892–1964

The Birth of the Universe

In the beginning, say most scientists, there was nothing. Then, 13 billion to 15 billion years ago, time, matter, space—everything—came into being in a cataclysm we call the Big Bang.

How It All Began
The universe began with a bang, and in one-millionth of a second was a mass of subatomic particles. Grouping together, these particles formed the first atoms of hydrogen and helium. After several billion years, gravity caused clouds of gas to collapse, forming stars and galaxies. The present-day universe is still expanding.

In 1924, Edwin Hubble turned the astronomical world on its ear when he announced that galaxies appeared to be moving concertedly away from Earth. What is more, Hubble discovered that the more distant the galaxy, the faster it was receding. The universe appeared to be blowing up, "inflating," like a balloon. The implications were astonishing: for the entire universe to be expanding, a propulsive force of unimaginable magnitude must have set matter on its outward-flowing course. The name astronomers eventually coined for this violent genesis was the Big Bang.

IN THE BEGINNING
Unfortunately, the term "Big Bang" is a misnomer that has led to some confusion. There was no "bang," no explosion in the strictest definition of the word, where one thing erupts into the space of something else. This is because there was no space. Instead, the Big Bang was more an unfolding of space and matter—from a region no larger than the period at the end of this sentence. This happened between 13 billion and 15 billion years ago.

A fraction of a second after the Big Bang, the universe was a hot,

First stars and galaxies form

Time and space begin with the Big Bang

Matter (electrons, protons, and neutrons) forms

Atoms (hydrogen and helium) form

eething mass of radiation and xotic particles. This expanded nd cooled, and more familiar orts of particles formed, including the neutrons, electrons, and rotons that make up everyday natter. Gradually, elements ormed—mostly hydrogen and elium—and these eventually collapsed under the influence of gravity to create galaxies, stars, nd planets.

What came before the Big Bang? Because time itself began t the Big Bang, there was no "before." The British astronomer Stephen Hawking once said that t would be like asking, "What is orth of the north pole?"

EVIDENCE FOR THE BIG BANG
Although still strictly a theory, he Big Bang has wide acceptnce because of three persuasive

HEARING THE ECHO OF THE BIG BANG

In 1965, Arno Penzias and Robert Wilson of Bell Telephone Labs were tracking sources of static in the receiver system of a radio telescope. Even after they had tightened the telescope's connections and removed some nesting pigeons from the antenna, there was one source of static that they were unable to eliminate. This turned out to be primordial background radiation coming from the entire sky. Further research revealed it to be the cosmic background radiation—the echo of the Big Bang itself. The discovery won Penzias and Wilson the 1978 Nobel Prize in Physics.

Today's observable universe has perhaps 50 billion galaxies

pieces of evidence: the redshift of the galaxies, the relative amounts of certain chemical elements in the universe, and the existence of cosmic background radiation.

When examined in a spectroscope, the spectral lines of nearly all galaxies appear shifted toward the red end of the spectrum, an indication that these galaxies are moving away from us (and from each other). It follows that in the past, galaxies must have been much closer and the universe much smaller.

All matter was created at the Big Bang, and like a cake, the universe today retains an imprint of its initial cosmic mixture. Analyses of the ratio of hydrogen to helium in the oldest stars agree exactly with what the Big Bang theory predicts should have been forged by that first explosion. Hydrogen accounts for roughly three-quarters of the total mass

and helium one-quarter, with deuterium, lithium, and heavier elements occurring in traces.

THE BIG BANG'S FAINT AFTERGLOW

Perhaps the most convincing single piece of evidence for the Big Bang is the cosmic background radiation (CBR), a uniformly distributed all-sky glow detected at the short end of the radio spectrum (see box, p. 207). It is thought to be the last vestige of heat from the Big Bang itself. After all this time, that initial outburst of fiery energy has cooled greatly, to −455°F (−270°C), just barely above absolute zero and some hundred million times cooler than a typical birthday candle.

In 1992, NASA scientists revealed that their Cosmic Background Explorer (COBE) satellite had detected slight temperature deviations in the CBR. From these small irregularities, which arose 300,000 years after the Big Bang, grew the clusters of galaxies, separated by vast voids, which we see today.

What COBE Saw
The universe is bathed in microwave radiation, the afterglow of the Big Bang. NASA's Cosmic Background Explorer (COBE) satellite (above) was launched specifically to study this radiation. The sky map (below right) produced by COBE shows ripples in the radiation—some hotter than average (pink/red) and some cooler (blue). The cold spots are parts of the early universe that were relatively dense—these are the "seeds" from which galaxy clusters grew.

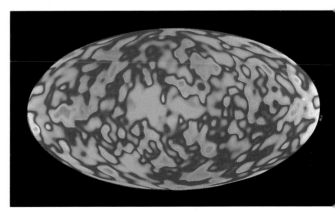

PRIMEVAL GALAXIES

When did galaxies first form? It was once thought that galaxies emerged relatively recently—two to three billion years after the Big Bang—when the universe was half as big as it is now. But recent discoveries have overturned this view and sent theorists back to the drawing board.

From 1996 through 2000, deep images of apparently empty regions of space made by the Hubble Space Telescope and Earth-based observatories have revealed countless faint blue galaxies in the remote universe. When astronomers examined the more luminous ones,

they realized they were looking at structures as they were at a time only a billion or so years after the Big Bang, when the universe was one-tenth its present age. Spectral analysis of light from these galaxies also revealed the presence of heavy elements, including carbon, silicon, aluminum, and sulfur, that did not exist when the universe began. Their presence means that at least one generation of stars must have already lived and died, dispersing these new elements made in their cores into the galaxy's interstellar medium, when it emitted the light that we see today.

Looking Back to the Big Bang

I n 1996, the Hubble Space Telescope used its privileged position in space to take a historic photograph of young, never-before-seen galaxies as distant as 10.5 billion light-years. Two hundred seventy-six exposures taken over 10 consecutive days were consolidated into one image (the Hubble Deep Field) showing objects four billion times fainter than can be seen by the human eye. Many of the galaxies are classical spirals and ellipticals while others display unusual shapes. Astronomers suspect that some of these galaxies began forming less than a billion years after the Big Bang. The region is a speck of sky in Ursa Major.

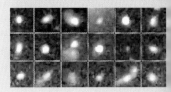

Dim Galaxies
Dim blue points of light in the upper Hubble image are shown in close-up in the lower one. Each of these 18 small galaxies is an incredible 11 billion light-years away.

Hubble Deep Field
Almost every point of light in this image is a galaxy—probably 2,000 all told. Some are so far away that we see them as they looked just after they formed 14 billion years ago—only about a billion years after the universe itself came into being.

The Farthest Galaxy
A small part of the Hubble Deep Field shows a very faint galaxy (arrowed) that may be more distant than any previously known. Scientists speculate that it formed only a few hundred million years after the Big Bang.

The Structure of the Universe

Astronomers know that the universe is filled with billions of galaxies arranged in immense clusters and superclusters. Now they must explain how such complexity evolved.

A Dark Matter First
Dark, unseen matter is thought to make up most of the universe, and U.S. astronomer Vera Rubin was the first to show its effects on individual galaxies.

Walls of Galaxies
On a very large scale, the universe has a distinctive structure of "walls" of galaxies separated by dark voids. The Great Wall is a sheet of galaxies stretching between the Coma and Hercules clusters.

On the very largest scales the universe is a strange place. Galaxies are clumped together to form sheets and filaments separated by vast voids, an arrangement that has been compared to a froth of soap bubbles or an elaborate honeycomb. Even stranger, we now think that most of the matter in the universe cannot be seen.

A SPACE ODDITY

Galactic surveys have revealed billions of galaxies strung out across walls and sheets and shot through with dark regions of apparently empty space.

The existence of such immense structures seems to conflict with accepted theories of how the universe began. The universe, say astronomers, started off in a smooth state—without large irregularities—and remained that way for at least 300,000 years. If matter and energy were once

distributed uniformly, how then did the present structure arise?

One theory is that 300,000 to 500,000 years after the Big Bang, random quantum motions made some regions of space denser than others. As the density of these regions increased further, their gravity also increased and they began to accumulate even more matter. In time, the dense regions collapsed into huge flattened structures. These primordial "pancakes" were oriented randomly with respect to each other and collapsed at different rates, thus setting up the honeycomb pattern seen today.

ON THE TRAIL OF DARK MATTER

Strange as it may seem, the stars, galaxies, and the other familiar objects in the night sky account for only a small part of the universe; invisible "dark matter" makes up the rest.

Astronomers came to this conclusion by measuring the motions of galaxies in clusters and superclusters. They found that individual galaxies are moving faster than can be explained by conventional theory and that some clusters are being tugged by something invisible that has a huge gravitational pull. These motions can be readily explained if a substantial amount of dark matter is factored in.

Void

Great Wall of galaxies

Void

450
350
Millions of light-years
150

Local supercluster

MACHO Approach

When a cluster of galaxies stands between Earth and more distant galaxies, the cluster's gravity bends light, producing a gravitational lens, such as this one. The more distant galaxies are magnified, and look like rings or arcs. Scientists study these lenses to try to find evidence of the gravitational effects of MACHOs.

Dark matter makes up at least 90 percent of the universe. The two main hypothetical candidates for this material are WIMPs (weakly interacting massive particles) and MACHOs (massive compact halo objects). WIMPs are somewhat heavier than neutrons but, as their name implies, they interact weakly with normal matter; MACHOs are huge, undetected populations of brown dwarfs (stars too small to shine), white dwarfs (faint stellar remnants), and black holes.

The nature of dark matter is elusive, and to date none has been directly observed. But the quest continues, in laboratories, underground mines, and particle accelerators around the world. One day astronomers may at last identify this strange material, the main component of the universe.

WIMP Hunt

Exactly how do you find WIMPs? These theoretical particles are not found in nature on Earth, but can they be detected in a laboratory? Using a particle accelerator, physicists at Fermilab, Chicago, are trying to do just that.

The Future of the Universe

What lies ahead for the universe? How and when will it all end? By applying the same fundamental physical laws used to understand how the universe began, astronomers are beginning to formulate answers to these questions.

The universe has been expanding since the Big Bang, but will it continue to do so forever? Astronomers have come up with a number of intriguing scenarios, ranging from never-ending expansion to a long return journey that ends with a Big Crunch. In all cases, whether expansion will change at some point in the future depends on the density of the universe, and that depends on the amount of matter it contains.

POSSIBLE FUTURES

If the density of the universe is greater than a certain critical value—that is, the universe contains a sufficient amount of matter—the force of gravity will at some point bring expansion to a halt and the universe will begin to collapse, culminating in a Big Crunch. Such a universe is called "closed." If the density of the universe is less than the critical value—there is not enough matter—expansion will continue indefinitely. This is termed an "open" universe.

There is another option. Much observational evidence suggests that the universe is "flat," precisely balanced between open and closed. That means the universe will expand forever, always decelerating, but never quite coming to a halt.

Yet another possibility is that expansion may actually accelerate. Supernovas observed in distant galaxies in

Big Crunch
The Big Crunch is one scenario for the fate of the universe. At the moment (top sphere), the universe is expanding. If there is enough matter in the universe, the expansion halts and swtiches into reverse. The end comes with an all-destroying crunch (bottom sphere).

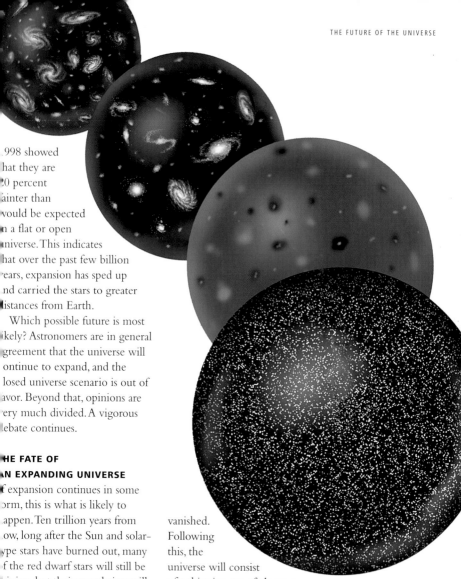

.998 showed
hat they are
20 percent
ainter than
would be expected
n a flat or open
universe. This indicates
hat over the past few billion
ears, expansion has sped up
nd carried the stars to greater
distances from Earth.

   Which possible future is most
kely? Astronomers are in general
greement that the universe will
ontinue to expand, and the
losed universe scenario is out of
avor. Beyond that, opinions are
ery much divided. A vigorous
ebate continues.

HE FATE OF
N EXPANDING UNIVERSE

f expansion continues in some
orm, this is what is likely to
appen. Ten trillion years from
ow, long after the Sun and solar-
ype stars have burned out, many
f the red dwarf stars will still be
hining, but their population will
ecline. A thousand trillion years
om now, the only remaining
ellar objects will be remnants
ke black holes, neutron stars, and
white and brown dwarfs.

   The black holes will outlive
verything else, but even they will
uccumb, diffusing into space as
hermal radiation and photons. By
0^{98} years (the figure 1 followed
y 98 zeros), they will have

vanished.
Following
this, the
universe will consist
of a thinning sea of electrons,
positrons, neutrinos, radiation,
and dark, empty space.

Indefinite Expansion
If the universe keeps
expanding, galaxies will
drift farther apart, stars
will eventually burn out,
and ordinary matter will
disintegrate, leaving a
boundless "sea" of
elementary particles.

ELEMENTARY PARTICLES
In an indefinitely expanding
universe, all matter will eventually give way to its simplest
constituents—elementary
particles, such as electrons,
positrons, and neutrons.

215

Life in the Universe

Are we alone? Of all the big questions astronomers must grapple with, none is bigger than the inquiry into whether or not life exists elsewhere in the universe.

Earth's Emissaries
Voyagers 1 and 2 are Earth's emissaries to the universe. Now on their way out of the Solar System, each carries a plaque showing Earth's location, and this disk, 'Sounds of Earth,' with recordings of human voices and music.

The possibility of life elsewhere in the universe has long fascinated people. During the 1st century BC, the Roman poet and philosopher Lucretius argued that "we must acknowledge that such combinations of other atoms happen elsewhere in the universe to make worlds such as this one... with races of different men and different animals..." In modern times, countless novels, movies, and TV series have found commercial success by depicting aliens and their strange worlds.

We need to feel that we are not alone, it seems, and that distance is all that prevents us from making contact. So, what are the possibilities?

LIFE IN THE SOLAR SYSTEM

There is no conclusive evidence to suggest that life exists in our neck of the cosmic woods. And yet spaceprobes have made some tantalizing findings.

Images returned in 2000 by NASA's Mars Global Surveyor spacecraft show what look like gullies formed by flowing water as well as deposits of soil and rocks transported by the flows. Scientists think these features were created by groundwater, similar to an aquifer. If life did develop on Mars, these "wet-

lands" would be prime places to look. A similar watery world may exist on the Jovian moon Europa beneath a layer of ice.

Scientists also want to probe the atmosphere of Titan, Saturn's largest satellite. A variety of organic compounds have been detected in Titan's thick atmosphere, including carbon monoxide and hydrocarbons such as hydrogen cyanide. This latter molecule is of particular interest since it is the basis for some of the components of DNA.

Sometimes evidence of possible extraterrestrial life is found much closer to hand. In 1996, scientists were studying a Martian meteorite that fell in

UFO Alert
Are these really flying saucers? Or merely odd-shaped clouds? Whatever the truth about this photo taken in Sicily in 1954, the large number of sightings of "alien spacecraft" over the years says a lot about our desire to believe that we are not alone.

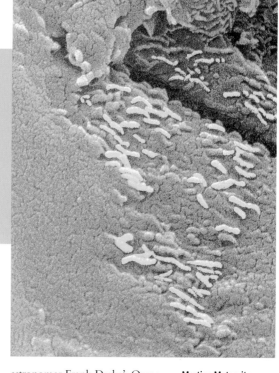

Antarctica 13,000 years ago. They discovered what appeared to be fossilized bacteria that may have originated on the Red Planet's ancient surface. Not everyone was convinced, however. Subsequent analysis by other scientists suggested that the tube-shaped structures may be the remains of organic compounds that entered the rock while it lay in the icy Antarctic wasteland. The debate continues.

SIGNAL SEEKERS AND INTERSTELLAR MESSENGERS
Astronomers began their search for extraterrestrial intelligence (SETI) in earnest in 1960. The astronomer Frank Drake's Ozma Project used a radio telescope to try to detect unusual signals from two nearby stars. The targets were Tau Ceti and Epsilon Eridani, which were considered sufficently Sun-like to be possible centers of planetary systems. Nothing out of the ordinary was detected.

The 1970s saw several attempts to let any intelligent life forms know about us. The Pioneer 10

Martian Meteorite
A small chunk of Mars fell on Antarctica about 13,000 years ago, to be discovered by scientists in 1984. Known as ALH84001, the meteorite drew worldwide attention in 1996 when researchers announced that it contained evidence of early Martian life. The evidence included tubular microstructures (colored yellow in the image above) that resemble bacteria, and the presence of organic compounds in globules (left) in the rock. Most researchers now doubt these are Martian microfossils. But that, they say, is no reason to stop looking.

Is Anyone Out There?
In the search for extra-terrestrial life, radio telescopes are normally used to pick up possible alien signals. But in 1974, the Arecibo radio dish in Puerto Rico was used to transmit a message about us. The message will take about 25,000 years to reach its target, the globular cluster M 13 in the constellation Hercules.

and 11 spacecraft each carried out of the Solar System a plaque engraved with male and female forms and a map showing the location of the Solar System. The later Voyager 1 and 2 spacecraft went one step further: they had a gold-coated copper audio-video disk which featured selections of music from around the world, as well as sounds of life on Earth.

In 1974, a complex message written in binary code was transmitted from the giant radio telescope in Arecibo, Puerto Rico, toward the globular cluster M 13. The message included the chemical formulas for the molecular components of DNA. In effect, we were telling possible alien listeners what to expect if they dropped in for a visit.

SETI COMES OF AGE

NASA began a comprehensive search for extraterrestrial radio messages in 1992. The nonprofit SETI Institute took over three

SERENDIP WANTS YOUR COMPUTER

SERENDIP, a SETI program conducted by the University of California, Berkeley, enlists the aid of thousands of private citizens and their personal computers. Called SETI@home, it provides computer owners with free, downloadable software screensaver programs. When the computer is idle, it automatically downloads 300 kilobytes of SERENDIP data for analysis. The results are ultimately sent back to the SERENDIP team and combined with the analyzed data from thousands of other SETI@home participants. Any intriguing signals are followed up at a later date.

ears later and renamed the earch Project Phoenix.

Using radio telescopes and omputers, Project Phoenix sifts hrough the countless natural ignals from space in an attempt o find signals that could only ave been generated artificially. Tens of millions of channels can e monitored concurrently. The gnals are filtered through software which sorts them by their oherency, such as pulses or epetitive signals. If unusual ignals persist, the search program lerts astronomers automatically.

Although no artificial signals ave yet been found, the project as plenty more raw material on vhich to draw. It aims to search hrough two billion channels for thousand stars within a radius of 200 light-years. Candidate stars nust be at least three billion years old and similar to the Sun, since hey are most likely to have Earth-like planets that may upport life.

S EARTH UNIQUE?

Superficially it seems a safe bet to assume that somewhere out there re worlds containing some sort of life. For one thing, rapidly ccumulating evidence shows that he universe abounds with other planetary systems (see p. 188), and vhere there are planets, there may be life. We also know that organic molecules—among the building blocks of living things—are very ommon in the universe.

However, some scientists consider life, particularly intelligent ife, an extremely rare occurrence. For them, a unique convergence

of factors made life possible here. Earth's distance from the Sun ensures that water remains in a liquid state, not ice or vapor. Our planet follows a stable orbit around a stable star and has a single large Moon that prevents the poles from tilting too far, thus preventing disastrous fluctuations in climate. Our atmosphere contains enough carbon dioxide to sustain life but not too much to create a runaway greenhouse effect, as on Venus. Finally, Earth is located in a quiet backwater of the Galaxy, far from massive exploding stars or energetic star-forming regions that could irradiate developing life forms. These charmed conditions allowed simple cellular forms to gain a foothold and then nurtured and sustained them so that more advanced life could slowly evolve.

Earth may be one-of-a-kind.

A Lonely Cosmos?
In spite of the earnest hopes and imaginative leaps of science-fiction writers and illustrators, Earth may very well be the only place in the universe where intelligent life exists. This fanciful magazine illustration of Jovian life dates from the 1930s.

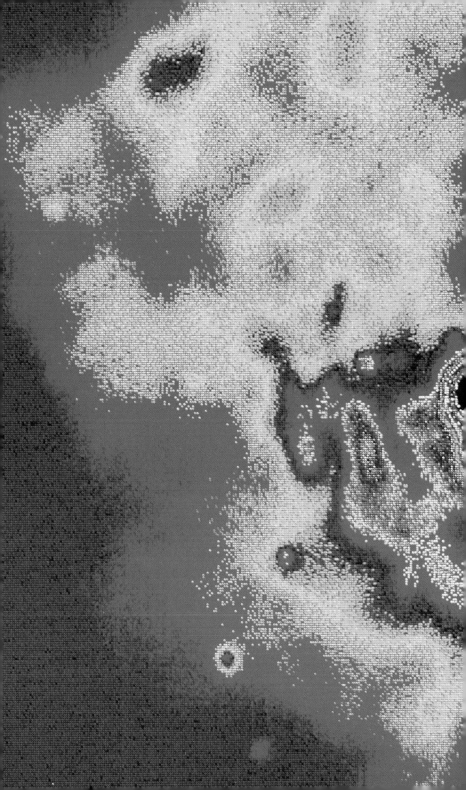

A Guide to

Celestial

Objects

How to Use This Guide

This is your field guide to outer space. Its detailed information will allow you to explore a wonderful variety of celestial objects from the Moon, Sun, and planets to stars, galaxies, and nebulas.

The Guide begins by looking at the two most familiar celestial objects—the Moon and Sun—progresses to the planets and our other neighbors (including meteors and comets), and then takes off for more distant realms.

The star chart chapters contain two types of charts: the Monthly Star Charts give you the big picture, showing the constellations month by month for both hemispheres; the Sky Tours contain much more detail, zooming in on deep-sky objects of interest.

Building on your star-chart experience, the final three chapters focus on deep space: stars, nebulas, and galaxies.

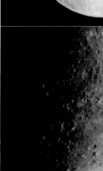

MONTHLY STAR CHARTS

Southern Hemisphere
January ▪ Looking North

OBSERVING THE MOON AND SUN

Observing the Moon

- Visibility: most surface detail seen at local lunar sunrise or sunset
- Apparent size: about 32 arcminutes
- Apparent magnitude: -12.7 (Full Moon); -10.5 (Quarter Moon)
- Diameter: 2,160 miles (3,476 km)
- Surface shaped by impacts and volcanism

The best time to look for details with binoculars or telescopes is around local sunrise or sunset on the Moon. This is when sunlight falls at a shallow angle on the ground, and the line between lunar night and day throws craters, domes, and hills into starkest relief. The flat lighting of Full Moon is best for tracing the bright rays and for examining varied shades that mark different lava flows in the dark "seas." (Early astronomers mistook these for dried-up ocean beds and called them *maria*, from the Latin word for sea.)

Whether you use binoculars or a telescope, make sure the optics are mounted steadily. Telescopes that shimmy and shake cannot show anything clearly. The quality of atmospheric seeing determines how much you can see on a given night, so do not observe over rooftops or parking lots, where rising warm air disturbs the view.

Start with the lowest magnification eyepiece your telescope has. Then slowly increase the power as one feature or another catches your eye. If views become unsteady, reduce the magnification until details sharpen again. (Viewing the entire Full Moon is almost painful, so bright is its disk, and some telescopic observers use gray filters to reduce the glare.)

The maps on pp. 229 and 231 have north at the top, showing the Moon as observers in the Northern Hemisphere see it with the naked eye or binoculars. Astronomical telescopes generally invert this view, putting south at the top. (Southern observers thus will find the maps likely match their telescopic views.)

Full Moon's Glare
At Full Moon, light from the Sun flattens all perspective, leaving only variations of light and dark caused by differences in age and composition.

Viewing Times
The best times to observe the Moon are when sunlight strikes it at a shallow angle, throwing features into sharp relief. This happens at First and Last Quarter Phases.

Lunar Portrait
Photographs taken with even small telescopes show an amazing amount of surface detail.

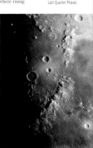

226

This illustration gives a good idea of how the object will look in your scope's eyepiece.

A bulleted list of key facts provides at-a-glance information about the object, such as size, visibility, and magnitude.

The text provides detailed information on various aspects of the object, such as how to locate it, what it should look like, and the best ways to view it.

The North Polar Sky

This northern sky tour focuses mainly on the constellations Ursa Major and Ursa Minor, and offers skywatchers plenty of celestial treasures. In Greek mythology, Ursa Major and Ursa Minor are the beautiful Callisto and her son Arcas, changed into bears by a jealous Hera, then flung into the heavens and immortalized by Hera's husband—and Callisto's lover—Zeus, chief of the gods.

Ursa Major

1 Polaris, The North Star (Alpha [α] Ursae Minoris)
Polaris is a Cepheid variable star 46 times larger than our Sun and 432 light-years away. Through a telescope, look for a faint (magnitude 9.0) pale blue companion star 18 arcseconds away from Polaris.

2 Mizar and Alcor (Zeta [ζ] and 80 Ursae Majoris)
Look at Mizar, the middle star in the Big Dipper's handle, with your unaided eyes. You should be able to make out a faint star, Alcor, just above it. A pair of binoculars splits them

Seen best in
Northern Hemisphere spring
Southern Hemisphere *
* not visible south of 10°S latitude

Sky Tour 1

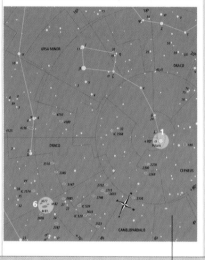

319

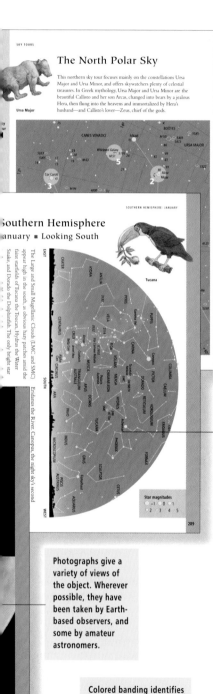

Southern Hemisphere
January ■ Looking South

The Large and Small Magellanic Clouds (LMC and SMC) appear high in the south, as obvious hazy patches amid the faint starfields of Tucana the Toucan, Hydrus the Water Snake, and Dorado the Dolphinfish. The only bright star

Tucana

Eridanus the River, Canopus, the night sky's second

Star magnitudes
-1 0 1
2 3 4 5

289

The Monthly Star Charts will help you find your way among the stars. Select from the 24 charts for each hemisphere.

The Sky Tours use the technique of star-hopping to let you zoom in on some of the most fascinating objects in the sky.

Photographs give a variety of views of the object. Wherever possible, they have been taken by Earth-based observers, and some by amateur astronomers.

Colored banding identifies the chapter—see the key on the right.

227

CHAPTER BANDING KEY

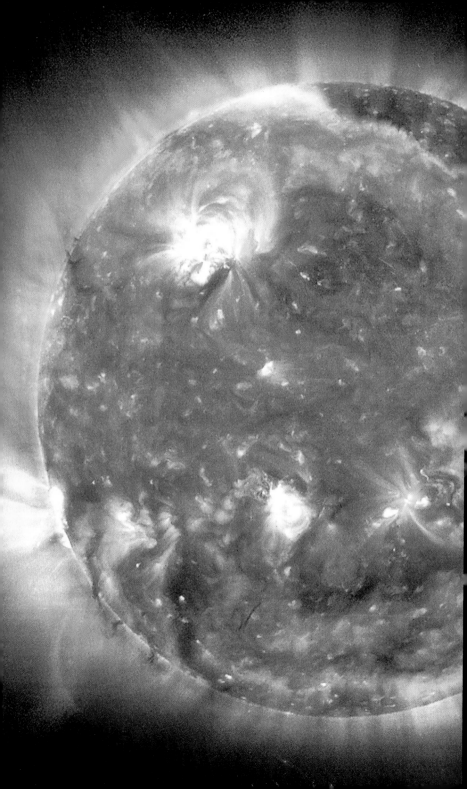

Observing the
Moon and Sun

The Sun and Moon were probably
the very first objects observed by the
earliest skywatchers. That age-old
fascination still runs strong, and
today these two beacons in the sky
are usually the first objects viewed
with a brand-new telescope. Each
has much to offer.

*"And God made two great lights:
the greater light to rule the day, and the
lesser light to rule the night..."*

Genesis 1:16

Observing the Moon

- Visibility: most surface detail seen at local lunar sunrise or sunset
- Apparent size: about 32 arcminutes
- Apparent magnitude: –12.7 (Full Moon); –10.5 (Quarter Moon)
- Diameter: 2,160 miles (3,476 km)
- Surface shaped by impacts and volcanism

The best time to look for details with binoculars or telescopes is around local sunrise or sunset on the Moon. This is when sunlight falls at a shallow angle on the ground, and the line between lunar night and day throws craters, domes, and hills into starkest relief. The flat lighting of Full Moon is best for tracing the bright rays and for examining varied shades that mark different lava flows in the dark "seas." (Early astronomers mistook these for dried-up ocean beds and called them *maria*, from the Latin word for sea.)

Whether you use binoculars or a telescope, make sure the optics are mounted steadily. Telescopes that shimmy and shake cannot show anything clearly. The quality of atmospheric seeing determines how much you can see on a given night, so do not observe over rooftops or parking lots, where rising warm air disturbs the view.

Start with the lowest magnification eyepiece your telescope has. Then slowly increase the power as one feature or another catches your eye. If views become unsteady, reduce the magnification until details sharpen again. (Viewing the entire Full Moon is almost painful, so bright is its disk, and some telescopic observers use gray filters to reduce the glare.)

The maps on pp. 229 and 231 have north at the top, showing the Moon as observers in the Northern Hemisphere see it with the naked eye or binoculars. Astronomical telescopes generally invert this view, putting south at the top. (Southern observers thus will find the maps likely match their telescopic views.)

Full Moon's Glar
At Full Moon, light fro the Sun flattens perspective, leaving or variations of light and da caused by differences age and compositio

Viewing Time
The best times to obser the Moon are when su light strikes it at a shallo angle, throwing featur into sharp relief. Th happens at First ar Last Quarter Phase

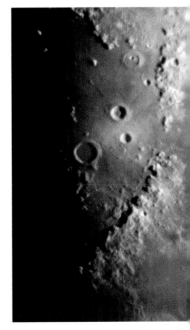

Lunar Portrait
Photographs taken with even small telescopes show an amazing amount of surface detail.

First Quarter Moon

- Visibility: visible in late afternoon and evening
- Contains sites of first and last lunar landings
- Has best example of an impact basin: Mare Nectaris
- Has best example of a wrinkle ridge: Serpentine Ridge
- Has largest expanse of ancient terrain: the highlands

The Moon's eastern hemisphere contains several lava-filled impact basins. Among them is Mare Nectaris, which was flooded only partly and which still shows portions of its upthrown basin rim, the Altai Scarp (Rupes Altai). Notice how features touched by the lava appear softened and melted. Mare Nectaris, 540 miles (860 km) across, lies on the edge of the highlands, an ancient terrain that also covers most of the Moon's far side. The highlands are so heavily battered that any new impact would destroy existing craters.

Mare Serenitatis, 570 miles (920 km) across, is another impact basin, but completely filled with lava. Across it snakes the Serpentine Ridge, the Moon's largest wrinkle ridge. Some 300 to 600 feet (100 to 200 m) high, it formed as the weight of the lava buckled the basin floor. The lava filling Serenitatis has different colors, showing that it came from separate eruptions. The light streak across the mare is a ray from the crater Tycho, hundreds of miles to the southwest.

Mare Crisium, 460 miles (740 km) across, appears elongated north–south thanks to perspective, but it actually extends east–west. This impact basin appears full of lava, while traces of a broken outer rim show that the Crisium basin, when fresh, reached twice its current size.

The first Moon landing (Apollo 11 in 1969) was in Mare Tranquillitatis, and the last (Apollo 17 in 1972) was in the Taurus Mountains (Montes Taurus) on the southeast shore of Serenitatis. Unfortunately, the Apollo landers, flags, and footprints are too small to be seen from Earth.

Sea View
Five "seas" occupy much of the First Quarter Moon: Mare Crisium, Serenitatis, Tranquillitatis, Fecunditatis, and Nectaris. Highlands dominate the south.

Lunar features
1 Albateginus
2 Aristoteles
3 Atlas
4 Endymion
5 Hercules
6 Hipparchus
7 Langrenus
8 Mare Crisium
9 Mare Fecunditatis
10 Mare Frigoris
11 Mare Nectaris
12 Mare Serenitatis
13 Mare Smythii
14 Mare Tranquillitatis
15 Mare Vaporum
16 Maurolycus
17 Montes Caucasus
18 Montes Taurus
19 Posidonius
20 Rupes Altai
21 Serpentine Ridge
22 Theophilus
23 Werner

24 Apollo 11 site
25 Apollo 16 site
26 Apollo 17 site

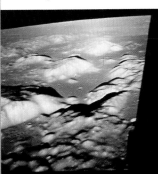

Apollo 17
In this NASA image, rounded hills flank the valley where Apollo 17 landed. In the distance lies Mare Serenitatis.

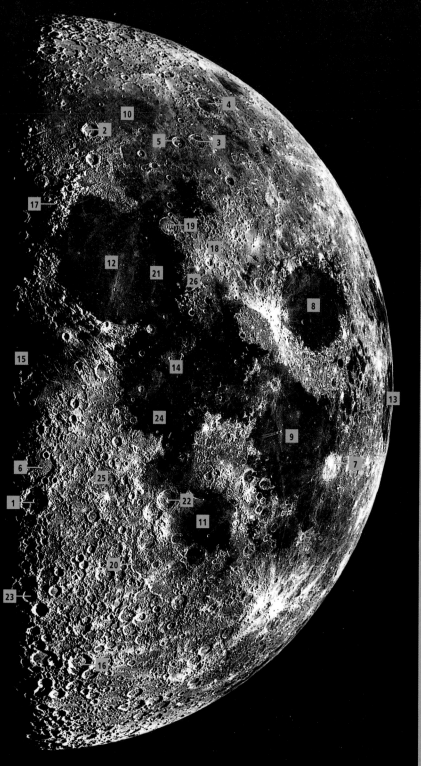

Last Quarter Moon

- Visibility: visible from after midnight until after sunrise
- Has Moon's most spectacular crater: Copernicus
- Shows best ray pattern: Tycho
- Has Moon's most dramatic mountain range: Montes Apenninus
- Has largest amount of lava coverage

The western hemisphere of the Moon contains Oceanus Procellarum, a vast lava flow connecting several lunar "seas." Nearby Copernicus is the Moon's most outstanding impact crater. It is 58 miles (93 km) in diameter, with gullies on its outer walls, collapsed terraces inside, and multiple central peaks made from material dredged up from deep within the Moon. Surrounding Copernicus is a splash of rays made of rock shattered when the crater formed about 800 million years ago.

North of Copernicus lies Mare Imbrium, some 720 miles (1,160 km) across. The impact creating the Imbrium basin scarred much of the Moon's face, 3.84 billion years ago. The lavas filling the basin are hundreds of millions of years younger. The impressive Apennine Mountains line part of Mare Imbrium with peaks over 16,000 feet (5,000 m) high. Apollo 15 landed at their foot in 1971.

Montes Alpes, the Alps, cut through by the Alpine Valley, contain the crater Plato, 63 miles (101 km) wide. Its interior was flooded by lava. To its west lies Sinus Iridum, a 160 mile (260 km) impact crater that lost part of its rim when lava filled the Imbrium basin.

Mare Humorum partly resembles Mare Nectaris, with partially destroyed craters. It is 260 miles (425 km) in diameter. Its best feature is a ruined crater, Gassendi, 68 miles (110 km) across. Gassendi was invaded by lava, and broken faults scar its floor. In the highlands to the south-east lies Tycho, 53 miles (85 km) in diameter, hub of the Moon's grandest ray system.

Crater Showpiece
The Last Quarter Moon's showpiece is the crater Copernicus, at the center of a system of rays

Lunar features
1 Albateginus
2 Alphonsus
3 Archimedes
4 Aristarchus
5 Clavius
6 Copernicus
7 Eratosthenes
8 Gassendi
9 Grimaldi
10 Hipparchus
11 Kepler
12 Mare Cognitum
13 Mare Frigoris
14 Mare Humorum
15 Mare Imbrium
16 Mare Nubium
17 Mare Vaporum
18 Montes Alpes
19 Montes Apenninus
20 Oceanus Procellarum
21 Plato
22 Ptolemaeus
23 Rupes Recta
24 Sinus Iridum
25 Tycho

26 Apollo 12 site
27 Apollo 14 site
28 Apollo 15 site

Two Craters
This view from Apollo 12 shows the crater Reinhold in the foreground and Copernicus on the horizon.

Observing Lunar Eclipses

- Visibility: visible over wide areas; no travel needed
- Optical equipment helpful but not necessary
- Occur twice a year on average
- Total phase of eclipse lasts over an hour
- Easy to see and photograph

A lunar eclipse happens when the Moon passes through Earth's shadow, which can occur only at Full Moon. The shadow extends behind Earth like a cone of darkness pointing at the stars. Lunar eclipses are either total or partial, depending on whether the Moon goes completely into the shadow. But not every Full Moon sees a lunar eclipse because the Moon's orbit tilts a few degrees relative to Earth's orbit around the Sun. This means that at most Full Moons, the Moon misses Earth's shadow, passing above or below it.

During a total eclipse, the Moon turns reddish-copper in color as it is lit by sunlight filtering into Earth's shadow. Depending on how dusty the atmosphere is, the eclipsed Moon may be bright or dark. Lunar eclipses are widely visible (unlike total solar ones): everyone on the night side of Earth while the eclipse is happening can enjoy it, if skies are clear.

Viewing by naked eye is easy, and observers with telescopes and binoculars often watch the moving edge of shadow darken craters. Others put a camera on a tripod and shoot scenes with an eerie, reddish Moon in the sky. Some photographers attach the camera to a telescope tracking the stars and make a multiple exposure. This captures the Moon as it moves through eclipse.

Alternatively, you can frame the Moon in the sky with a foreground tree or building, and then take a series of exposures on the same frame exactly 5 or 10 minutes apart. ISO 400 film captures partial phases with a 1/125th second at f/8; during totality use 1 second at f/4. With other film speeds, adjust appropriately.

Time Laps
The photographer of th image placed the partial eclipsed Moon at the uppe left of the frame ar snapped a short exposure Then, without changin the camera's aim advancing the film, he too an exposure every fe minutes as the Moo slipped into total eclips and emerged agai

Ruddy Moon
In a total eclipse, sunlight refracting through Earth's atmosphere stains the gray lunar dust with the hues of sunrise and sunset.

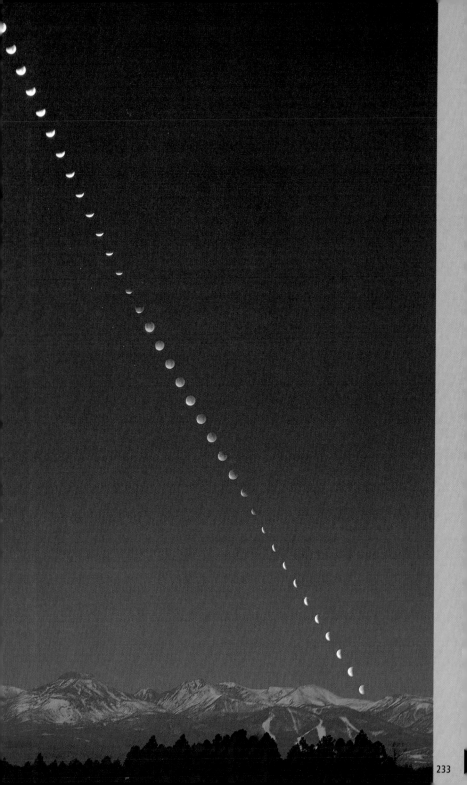

Observing the Sun

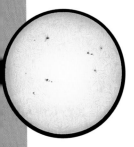

- Never look at the Sun through binoculars or an unfiltered telescope
- Visibility: sunspots are seen best at maximum solar activity
- Apparent magnitude: –26.7
- Apparent size: about 31 arcminutes
- Diameter: 865,000 miles (1,392,000 km)

Always look at the Sun with proper protection—you could be blinded by even a split second of unfiltered light. To view the Sun safely, use a solar filter that fits over the telescope's aperture (solar filters that fit over the telescope's eyepiece are not safe and should be avoided). Cover the finderscope to prevent scorching someone or something by accident. An alternative to using a proper solar filter is to set up your telescope with a Sun projection screen (see illustration below).

The Sun's surface, the photosphere, shows many telescopic features. At high magnification it has a granular appearance, like oatmeal. The granules are cells of hot rising gas, with the smallest spanning 700 miles (1,100 km). Large, irregular bright patches called faculae show up best near the Sun's edge, which looks slightly darker.

The main features, however, are sunspots. These are places where the Sun's magnetic field becomes twisted enough to block the normal outflow of heat. Therefore sunspots are cooler places, hence darker. On rare occasions a sunspot may become big enough to see by eye (look through a solar filter only). Sunspots are intensely magnetic, and wax and wane over an 11-year cycle. The last sunspot maximum occurred in 2000–2001.

Sometimes magnetic activity erupts as flares, which shoot charged particles into space. Powerful flares can disturb Earth's ionosphere and cause magnetic storms and auroras.

The corona, the Sun's outer atmosphere, has a temperature of millions of degrees but is so dim that it can be seen only during a total eclipse.

An ordinary solar filter shows the "white-light," everyday Sun. To see more solar activity, some observers buy special filters that isolate a narrow wavelength, usually hydrogen–alpha (6563 Å). These can reveal prominences and other surface details.

Darkness Descends
Sunset ends another day's solar observations. Many people find that solar-viewing fits more easily into their daily lives than does nighttime astronomy.

Spots on the Sun
The Sun almost always displays a few sunspots. Observations of sunspots show the Sun rotates about once a month.

Projecting the Sun
To set up your telescope to project the Sun, first cap the finderscope, then position the telescope by moving it until its shadow is at a minimum. The Sun should be shining through the telescope and out through the eyepiece, toward the ground. Then project the Sun's image onto a piece of paper, shielding this screen from direct sun. Lastly, focus the eyepiece until the image is sharp.

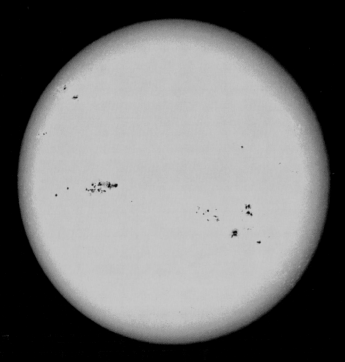

Observing Solar Eclipses

- Partial solar eclipses each year (on average): 3
- Total solar eclipses each year (on average): 1
- Maximum length of total eclipse: 7 minutes 40 seconds
- Total eclipses visible only along a narrow path
- Total eclipses offer best chance to see the solar corona

Solar eclipses can be total, partial, or annular. In a total solar eclipse the Moon covers the whole disk of the Sun, leaving only its corona and a few prominences uneclipsed. A partial eclipse leaves some of the Sun's surface uncovered, while an annular eclipse is a dramatic type of partial eclipse. In it, the Moon passes directly over the Sun, but appears too small to eclipse it entirely. Thus the Moon is encircled by a bright ring—an annulus—of uneclipsed Sun.

Observing a solar eclipse is much like ordinary solar observing. During a partial eclipse (or during all partial phases of a total eclipse), use a Sun filter or a projection screen (see p. 234) to protect your eyes and telescope. This lets you watch the lunar edge crawl across the Sun, engulfing sunspots. As sunlight diminishes, do not be tempted to look at the Sun with the naked eye. If an eclipse is total, once the Sun is completely eclipsed, you can remove the filter and safely view without it.

As the total phase begins, the last rays of sunlight stream through valleys on the Moon's edge, a phenomenon called Baily's beads. A soft halo of pearly light surrounds the black disk of the Moon. This is the corona, the Sun's extremely hot outer atmosphere. It is too dim to see except during an eclipse. A telescope may also show solar prominences reaching up from behind the Moon. Totality ends in a burst of light, so keep the solar filter ready.

Grand Event
A total solar eclipse is one of Nature's grandest spectacles, and is something everyone should try to see at least once in their life

Baily's Beads
Below left: Just before totality comes the shimmering Baily's beads phenomenon, in which sunlight sparkles through the valleys of the Moon.

Annular Eclipse
Below: In an annular solar eclipse, the Moon does not cover the Sun completely, and a ring of the Sun's uneclipsed surface surrounds the silhouetted lunar disk.

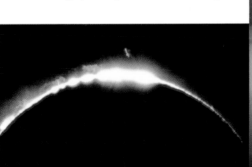

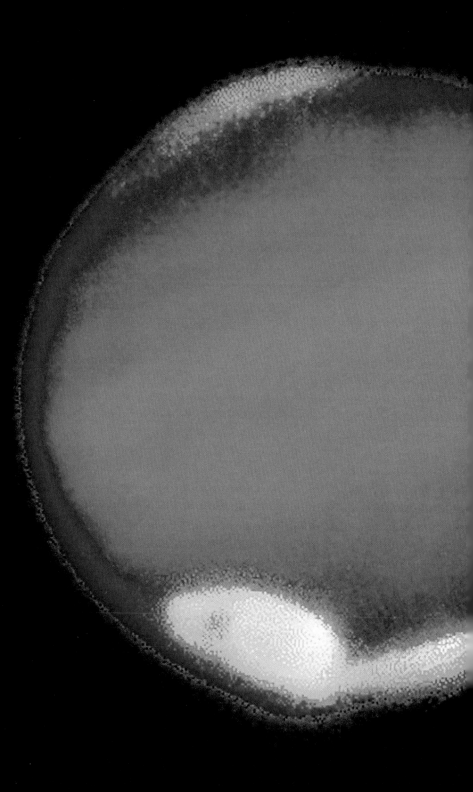

Observing
Planets and other Neighbors

The bright crescent of Venus; a dust storm on Mars; Jupiter's Great Red Spot; the incomparable rings of Saturn; the mesmerizing drama of a meteor shower—our neighbors in the Solar System offer all this and more to those who know how and where to look.

"In the field of observation, chance only favors those minds which have been prepared."

Louis Pasteur
French microbiologist, 1822–95

Observing Planets

- Visibility: generally visible every night in the year, although not all can be seen at once or on any given date
- Consult astronomy magazines, almanacs, or software for planet positions
- Inner planets: Mercury and Venus
- Outer planets: Mars, Jupiter, Saturn, Uranus, Neptune, and Pluto

Planets move slowly among the stars, following patterns that repeat over months or years. Two inner planets (Mercury and Venus) orbit closer to the Sun than Earth, while six outer planets (Mars through Pluto) orbit farther out than Earth. This difference determines how each planet moves during its apparition, or period of visibility.

Mercury (or Venus) first appears low in the western sky after sunset. Day by day it climbs higher, becoming most visible at greatest eastern elongation. It then passes between Earth and the Sun (called inferior conjunction) and moves into the morning sky for best visibility at greatest western elongation. It then moves onward to round the far side of the Sun. This is superior conjunction, the start of a new cycle.

An outer planet's apparition begins as it becomes visible low in the east just before sunrise. Each day the planet rises earlier. Eventually it rises at sunset, a point called opposition which marks best visibility. The planet then starts setting sooner and sooner after the Sun, and finally disappears into the solar glare. The process restarts when the planet emerges at dawn again.

Telescopes for planet watching need sturdy mountings and high-quality optics. Most observers use powers of 200x or less, due to unsettled seeing. Also useful are colored filters, which accentuate specific features on the planets.

Sky Lights
The two brightest objects in the night sky are the Moon and Venus.

Four-planet Line-up
Four of the five naked-eye planets lined up to get their picture taken in early March 1999. From the horizon upward, they are Mercury, Jupiter, Venus (the brightest), and Saturn. Such gatherings in the dawn or dusk are rare.

Interior–Exterior
Below: Interior planets such as Venus reach best visibility at the points of greatest elongation, while exterior planets such as Mars are best seen at opposition. All planets disappear into the Sun's glare at conjunction.

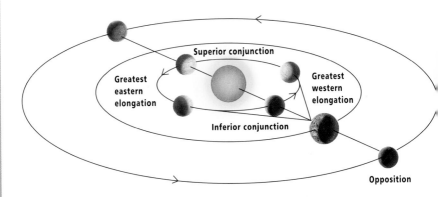

Observing Mercury

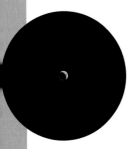

- Visibility: seen only in twilight, soon after sunset, or before sunrise
- Apparent magnitude: +3 to −2
- Apparent size: 5 to 13 arcseconds
- Diameter: 3,029 miles (4,875 km)
- Heavily cratered, lunar-like surface

Mercury is little seen. As the Solar System's innermost planet, its orbit keeps close to the Sun, so the planet is visible only shortly before sunrise or briefly after sunset. Best viewing times for Northern Hemisphere observers are evening apparitions in March–April and morning ones in September–October. (In the Southern Hemisphere, the best evening apparitions are in September–October, and morning ones in March–April.) During times of best visibility, Mercury stands low in the twilight for about three weeks. It looks like a warm-white star, and binoculars often help you spot it.

In a telescope Mercury's apparent size (and brightness) vary according to its distance from Earth and place in orbit, with the planet showing phases like our Moon. The phases can be hard to discern because Mercury's disk never appears large, and poor seeing near the horizon seldom leaves it steady. At moments of quiet seeing, however, the planet presents a grayish disk like a tiny Moon with no features. When seen near elongation, it appears half-lit, and close to the time of inferior conjunction, it shows a crescent.

On rare occasions, Mercury passes directly between Earth and the Sun. These events are known as transits. The next transits occur on May 7, 2003, November 8, 2006 and May 9, 2016. Each lasts about four hours. Prepare your telescope for solar viewing (see p. 234) and locate the tiny black dot moving slowly across the face of the Sun.

Mariner 10, the only spacecraft to visit Mercury, radioed back images of an airless, rocky surface covered with many craters and showing several giant impact basins. Unfortunately, none of this detail is visible to Earth-bound telescopes.

Mercury Occulted
The Moon occasionally covers, or occults, a planet or star. Consult astronomy magazines, almanacs, or sky-charting software programs to find out when occultations are going to happen, and whether they will be visible from your location. Here, Mercury is emerging from occultation behind the crescent Moon.

Spotting Mercury
Because it stays near the Sun, spotting Mercury means searching twilight skies for a dot of light. Binoculars often help. Here, Mercury is the "star" closest to the horizon, just below Jupiter. At such times, telescopic views are usually disappointing because of poor seeing.

Transit Viewing
A full-aperture filter ensures safe viewing of a transit of Mercury.

Observing Venus

- Visibility: best seen not long after sunset and before sunrise
- Apparent magnitude: –4.0 to –4.6
- Apparent size: 10 to 64 arcseconds
- Diameter: 7,521 miles (12,104 km)
- Lava plains and "continents" hidden under white clouds

Even novice stargazers will have no difficulty finding brilliant Venus, often called the morning or evening star. Venus appears bright enough to look like an airplane with landing lights on (or a UFO), and can even cast a faint shadow under the right conditions. It shines brightly because its thick white clouds reflect 76 percent of the sunlight falling on them.

Its cycle of appearances resembles Mercury's (see p. 242), but lasts longer and is easier to follow. Over eight years, Venus makes five apparitions. Its best visibility runs from before elongation in the evening sky, through inferior conjunction, until after morning elongation. This period lasts about 25 weeks. When Venus is near greatest elongation, its size in a telescope is about 25 arcseconds and its phase appears half-lit. Near inferior conjunction, Venus spans more than 60 arcseconds (1 arcminute) and shows a thin crescent phase.

Venus' brightness is dazzling against a dark sky. Therefore experienced observers view it in twilight, or even in daylight. They track the planet as dawn becomes day, or offset the telescope from the Sun.

In a telescope, Venus' opaque clouds usually appear feature-less. Sometimes, however, observers are able to make out faint, darkish patches on the white disk—probably cloud features. When the planet shows a crescent phase, some observers have reported seeing a faint glow on the dark hemisphere. Called the "ashen light," the glow is probably an optical illusion.

Venus occasionally passes across the Sun's face at inferior conjunction. Such events are called transits and they occur in pairs eight years apart. The next will occur on June 8, 2004 and June 6, 2012. In 2004, Venus crosses the Sun's southern part and in 2012, the northern.

Venus Occulted
Venus is occulted by a crescent Moon. An occultation is basically a form of eclipse.

Evening Sta
Venus can outshine everything in our view except the Sun and Moon Even twilight glow and cit lights do little to du the planet's light

Observing Mars

- Visibility: best seen six weeks before and after opposition
- Apparent magnitude: +2 to –2
- Apparent size: 4 to 25 arcseconds
- Diameter: 4,213 miles (6,780 km)
- Cold, cratered desert, very thin atmosphere, polar caps

Mars reaches opposition and best visibility every 780 days on average, about two years, seven weeks. At opposition its distance from Earth can vary from 35 million miles (56 million km) to 63 million miles (101 million km), producing nearly a twofold change in apparent size.

On first viewing Mars by telescope, beginners are often startled by its smallness. Even at its closest approach, Mars appears no larger than a small lunar crater. From Earth, the best telescopic views resemble naked-eye views of the Moon. Typical first looks show an amber disk, some faint markings, and maybe a whitish polar cap.

Mars is a planet with delicate, elusive features. To see it best, use at least a 5 inch (125 mm) refractor, a 6 inch (150 mm) reflector, or an 8 inch (250 mm) catadioptric scope. Filters held over the eyepiece also help: blue (Wratten 44a) and green (58) filters reveal clouds and atmospheric features, while orange (21 or 23a) and red (25) ones enhance surface markings.

Mars' four seasons each last about twice ours. The season on Mars at opposition is one season ahead of Earth's at that time. The polar caps grow and shrink, while dark markings, once believed to be vegetation, are vast lava flows. Their visibility changes as winds blow dust across them. Sometimes no markings appear for days or weeks, thanks to planet-wide dust storms. Observers should always be alert to their onset.

Polar Cap
When the view is calm and steady, and using high power, Mars' ice caps appear as whitish patches. You may be able to see one of the caps grow or shrink as a Martian season progresses. The southern cap is shown here.

On the Move
Mars can appear at any place in the sky on the ecliptic, rather than always staying near the Sun, like Mercury and Venus. Here it is, glowing dull orange, in Scorpius (facing page, on the right) and in Taurus (below).

Observing Jupiter

- Visibility: easy to spot, apparitions last about a year
- Apparent magnitude: –1.2 to –2.5
- Apparent size: 33 to 50 arcseconds
- Diameter: 88,730 miles (142,796 km)
- No solid surface, deep atmosphere with cloud belts

To the naked eye, Jupiter is a bright white "star" that slowly moves along the ecliptic, taking 12 years to circle the sky once. It is the largest of the gas giants and the most changeable planet visible in a telescope, displaying a constantly varying cloudscape as it rotates in less than 10 hours. Like Mars, Jupiter rewards patient observing.

Jupiter is mostly hydrogen and its "surface" is the cloudy top of a deep atmosphere, roiled by internal heat. Compounds of sulfur and phosphorus may produce the colors in the bands visible with a telescope. Jupiter's rapid rotation smears cloud features into east–west stripes paralleling the equator and also gives it a pronounced oval shape. The light-colored zones you see are high, cool clouds in regions of updrafts, while the darker belts mark warmer areas of downdrafts.

The north and south equatorial belts are the least variable, but all belts change in strength and position. They show subtle colors as well—browns, tans, yellows, oranges, and shades of blue-gray—which can be made to appear more pronounced with filters such as light blue (Kodak Wratten 80A or 82A) and yellow or orange (12 or 21).

Jupiter's most famous single feature is the Great Red Spot, a vast and turbulent eddy located just below the south equatorial belt that has been observed for at least 300 years. Its color and size change, perhaps in response to solar activity, being redder at solar maximum. When the Great Red Spot has faded, look for the Red Spot Hollow, an indentation that surrounds it.

Jupiter makes a good subject to draw. Before you begin, look at the planet for a few minutes to note the belts and zones. Once you start (using a 2B pencil), you will need to draw the major features in about 10 minutes, since Jupiter's rotation carries features along quickly. You can fill in the details later.

Changing Views
Jupiter's disk offers the backyard astronomer a wealth of ever-changing detail that can be seen in almost any telescope. Bright zones and dark belts vary in strength and change position slightly. The Great Red Spot fluctuates in size and color.

Lord of the Ecliptic
As Jupiter makes it way slowly along the ecliptic, it passes various clusters, nebulas, and bright stars. Here it lies near the Beehive Cluster in Cancer, creating a beautiful sight for binoculars.

Jupiter Occulted
Occultations of Jupiter tend to be more dramatic than those of other planets because Jupiter looms relatively large in the telescope's field of view.

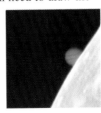

Observing Jupiter's Moons

- Visibility: four brightest are visible in binoculars
- Apparent magnitude: Io +5.0, Europa +5.3, Ganymede +4.6, Callisto +5.7
- Actual diameter: Io 2,263 miles (3,643 km), Europa 1,939 miles (3,120 km), Ganymede 3,273 miles (5,268 km), Callisto 2,983 miles (4,800 km)
- Biggest four discovered in 1610 by Galileo

Like a miniature Sun and planets, Jupiter controls a solar system of moons numbering at least 24. The four biggest and brightest are Io, Europa, Ganymede, and Callisto, in order from Jupiter outward. They were discovered in January 1610 by Galileo Galilei, in one of his first telescopic discoveries, and they are collectively named for him.

You can duplicate Galileo's feat with a pair of ordinary binoculars. Observe Jupiter at the same time each night. On a piece of paper plot a big dot for Jupiter and put smaller dots on either side to represent the minute points of light that flank it. In a week or so, the pattern of the moons' movements will emerge.

In a telescope, these moons are 5th and 6th magnitude in brightness, and resemble nothing more than points of light, even at high power. However, the plane of their orbits lies in Jupiter's equator and is close to our line of sight to the planet. Thus at times you can see one or more slowly cross Jupiter's disk over several hours. The moon may be hard to spot, but its dark shadow shows clearly.

On the far side of their Jovian orbits, the moons go behind Jupiter with an odd three-dimensional effect. Before Jupiter reaches opposition, moons enter the planet's shadow before going behind the planet, thus they seem to vanish in space. After opposition, the shadow lies differently, and moons reappear equally abruptly as they leave Jupiter's shadow.

Each satellite is distinctive. Io is fiercely volcanic, repaving itself with lava every few million years. Europa hides an ocean of water under a bright icy crust. Ganymede may also harbor water under a rock-and-ice surface, and it has a magnetic field. Callisto bears an ancient, heavily cratered crust.

Satellite's Transi
This sketch of Jupiter show one of the Galilean moor and its shadow, captured i transit across the disk

Mini-Solar Syster
Facing page, above: telescope or binoculars le you experience wha Galileo saw when h discovered Jupiter's fo largest moons—a sol system in miniatur

Shuttling Moor
Facing page, below: Th series, taken over fo successive nights, follov the Galilean moons as th shuttle around their hom planet. On one night, four moons may be visib on another you may s only two, with a third front of or behind Jupit

Observing Saturn

- Visibility: resembles a fairly bright star; telescopes show three rings
- Apparent magnitude: +3 to −2
- Apparent size: 15 to 21 arcseconds
- Diameter: 74,900 miles (120,500 km)
- Biggest and brightest ring system of any known planet

To the naked eye, Saturn is just a brightish "star" creeping along the ecliptic, taking 30 years to circle the sky. But when someone sees Saturn for the first time through a telescope, they almost always exclaim with delight—thanks to the rings.

Being a gas-giant planet like Jupiter, Saturn's "surface" is a cloud deck at the top of a deep atmosphere of hydrogen and helium. It displays fewer features than Jupiter, but slightly darker equatorial bands are visible in medium-size telescopes. On rare occasions, a large white cloud appears near the equator. To see Saturn's low-contrast cloud features better, try using an 80A or 82A blue filter or view in mid-twilight.

Almost any small telescope will show Titan, Saturn's largest moon, and at least three other moons will be apparent in a 6 inch (150 mm) telescope. Unlike the moons of Jupiter, which seem to always be in a line, Saturn's moons can be above or below the planet, because like Earth, the planet is tipped relative to the plane of the Solar System.

For many, however, the rings are the main attraction. Because Saturn's axis tilts 27 degrees to its orbit, our view of the rings is constantly changing as the planet moves around the Sun. At times the rings are tipped toward us, at others they are seen edge-on. In 2009, they will disappear from view for about a week, being too thin to see.

As seen from Earth, the rings have three bands. The outermost is the A ring. The dark Cassini Division separates it from the B ring, which is the widest and brightest. The gauzy C (or crepe) ring lies on the B ring's inner edge, and is hard to see except when the rings are tipped open.

Memorable Sight

Saturn is probably the most memorable sight in a telescope you can have. Visible in this photo are the dark Cassini Division in the rings and faint cloud markings.

Changing Aspect

As Saturn orbits the Sun once every 30 years, we see its rings slowly change appearance. When the rings are seen edge-on, which happens twice each Saturnian year, they briefly disappear from sight in small telescopes. These changes confused Galileo, the first to view Saturn by telescope. His instrument was not powerful enough to show the rings' true appearance, and he came to the conclusion that the planet had handles that came and went.

When to Look

Viewing Saturn in twilight can help reveal the planet's cloud features. In full darkness, try a filter.

Observing Meteors

- Visibility: best seen by naked eye; the greatest number of meteors is visible in the after-midnight hours
- Few meteors last more than a second or two
- Some major showers: the Perseids of August, the Leonids of November, the Geminids of December

Meteors or "shooting stars" are bright trails of light caused by bits of interplanetary debris entering Earth's atmosphere at speeds up to dozens of miles per second. Friction with air molecules makes them glow, then vaporize. Some meteors leave a train lingering a few seconds to a minute. Meteors as bright as Venus are called fireballs. Those that appear to break up are called bolides (pronounced "*bo*-lydes").

Observing requires little except warm clothing and a lawn chair, although binoculars can help reveal fainter meteors. You can see four or five meteors an hour on any moonless night. Such random meteors are called sporadics. They are more plentiful after midnight, when your viewing site turns onto the leading hemisphere of Earth as it travels in its orbit.

At certain times of year, Earth passes near a stream of cometary debris and a meteor shower occurs. This greatly increases the number of meteors. The comet debris is orbiting the Sun in parallel paths, so the meteors appear to come from one point in the sky, the radiant. Meteor showers are named for the constellation or star nearest the radiant. Several dozen major and minor showers occur each year.

The table on p. 149 shows the dates of peak activity for the major showers. Set up your lawn chair roughly facing the constellation the meteors are named for. (If you are not sure where that is, turn to the sky maps on pp. 260–311.) As your chosen constellation rises higher in the sky, you will see an increasing number of meteors— unless the night is bright with moonlight. This washes out most dim meteors, leaving only the brightest.

Vapor Tra

A rare daylight sight: vapor trail in the sky mac by a meteor as it vaporize in the upper atmospher Winds have distorted th once-straight trail in a smoky smea

Meteor Showe

Bright streaks against starry background a part of the 1998 Leon meteor shower (facir page, below). Because perspective, the meteo appear to radiate from or point in the sky. Mo showers are thus name for the constellation which this point (th radiant) lies. For th Leonids, the radiant is Leo; for the Geminic (below), it lies in Gemir

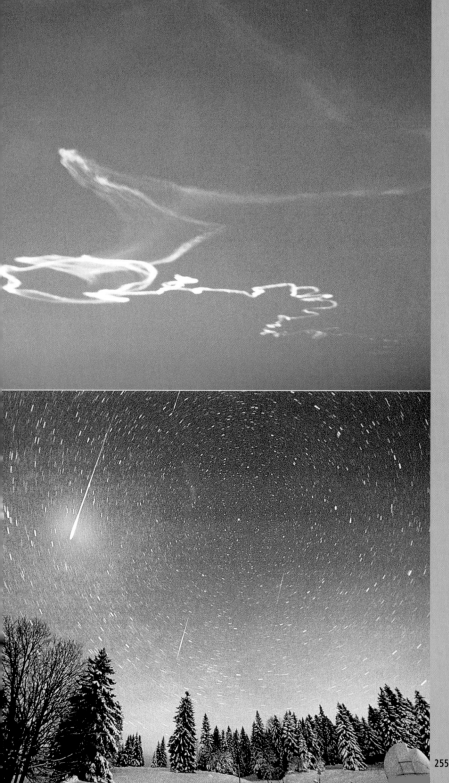

Observing Comets

- Visibility: naked eye to telescopic, apparitions last weeks to months
- Apparent size: a few arcminutes to many degrees
- Apparent brightess: barely visible to very conspicuous
- Diameter: nucleus only a few miles across; tail may reach hundreds of millions of miles

Unlike with planets and moons, the visibility of comets is unpredictable. Bright ones like Comet Hale-Bopp in 1997 and Comet Hyakutake a year earlier arrive unheralded, appearing out of the night to glide slowly across the heavens for a few magical weeks. When the next such "great comet" will arrive is unknown—we must simply wait and watch.

Comets are wonderful sights—by unaided eye, in binoculars, or through a telescope. The naked-eye view shows the entire comet, and observers frequently note that they display two distinct tails, a bluish one of ionized gas and a whitish one of dust. These tails sometimes point in slightly different directions. A pair of binoculars takes you closer in on the head and tail, and lets you see more details. These may include streamers or knots in the gas tail.

A telescope lets you study the comet's head far more closely. Many comets show a bright starlike point in the center of the head. But this is not the actual nucleus—the comet's icy core—which is too small to see from Earth. Instead, the bright center is the active cloud of gas and dust that surrounds the real nucleus and hides it from view. As you examine the comet's head closely, look for bright shells and streamers of light—these are produced by gas eruptions on the nucleus. As the icy nucleus spins, it shoots jets of gas that curve like the water from a whirling lawn sprinkler.

According to a tradition going back more than 200 years, new comets are named after their discoverers. Until a comet discovery is officially announced, up to three independent observers can earn the credit for finding it.

Decade of the Comet
The 1990s will long be remembered as the comet decade. Comet Levy led the way in 1990, then came Comet Shoemaker-Levy four years later. More impressive was Comet Hyakutake (above), which reached a maximum magnitude of –1 in March 1996. But the show stopper was Comet Hale-Bopp (right), in March–April 1997. Its brightness (magnitude –1.5) and distinctive forked tail made it the comet of the decade and one of the great comets of the 20th century.

Solar System Challenges

- Visibility: visible in binoculars, better seen in a telescope
- Uranus: blue-green, 6th magnitude, 3 or 4 arcsecond disk
- Neptune: blue-gray, 8th magnitude, 2 or 3 arcsecond disk
- Pluto: white, 14th magnitude, 0.1 arcsecond disk
- Asteroids: white, 8th magnitude and fainter, no visible disk

The Moon, Mars, and Jupiter look grand in any telescope. But the two outer gas-giant planets, Uranus and Neptune, and the enigmatic last planet, Pluto, pose tougher challenges. And the same holds for the fragmentary micro-worlds orbiting between Mars and Jupiter: the asteroids.

Uranus was discovered in 1781 by the English amateur astronomer William Herschel using a 6 inch (150 mm) telescope. It appears small, only 3 or 4 arcseconds across, and at 6th magnitude. To find it, check an astronomy magazine or software for its position. In the eyepiece it looks distinctly non-starlike and has a vivid blue-green color thanks to methane in its deep gaseous atmosphere.

Neptune was discovered in 1846, and looks like a dimmer, blue-gray echo of Uranus. It appears 8th magnitude and is only 2 to 3 arcseconds across.

It took a photographic search to find Pluto (in 1930). This planet, which is smaller than our own Moon, is 14th magnitude and shows no disk in backyard telescopes. Finding it requires accurate position information and a telescope of at least 10 inches (250 mm) aperture.

Asteroids orbit much closer to the Sun, but most are only a few miles across and faint. Several hundred lie within reach of a 3 inch (75 mm) telescope, however. Positions are available using software or the Internet. Locate the right field, plot all the stars you see, and recheck a day or two later. The "star" that moved is the asteroid.

Two Faint World
For beginners, findir Uranus (circled on left) ar Neptune (right) against background of stars no mean feat. Using higher magnification, th planets will show up tiny disks, although Jupiter-like surfa features will be visib

Asteroid Hunt
To hunt down an asteroid you will need a telescope, position information, star charts—and patience. The short streak at the center this photo is an asteroid.

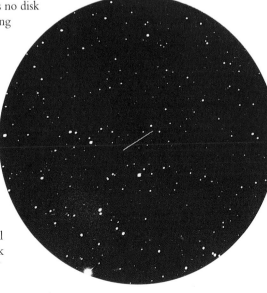

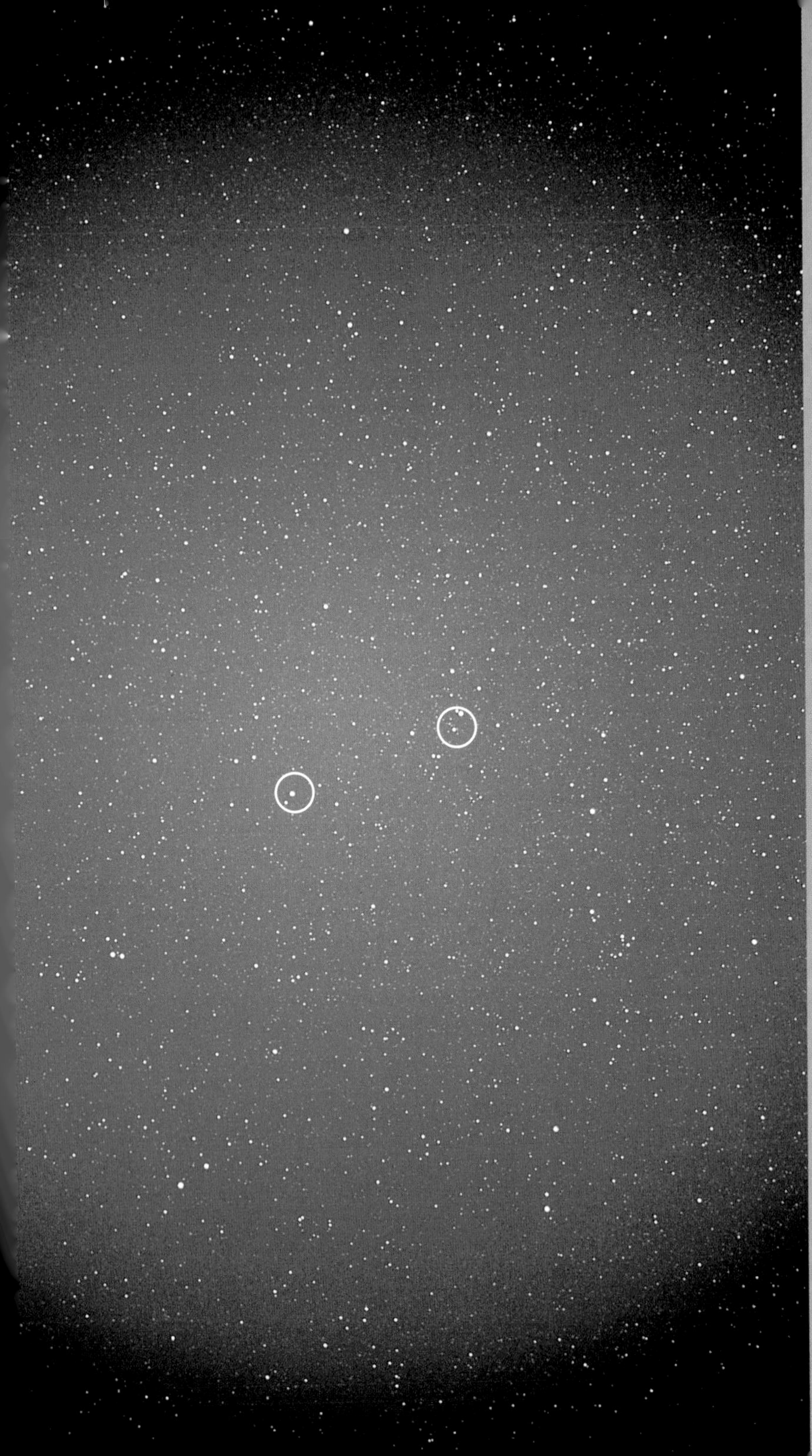

Monthly
Star Charts

Star charts may look chaotic or intimidating at first glance, but they are surprisingly easy to use. After all, they are simply maps of the night sky. As you gain experience in their use, you will find that they become as familiar to you as any roadmap.

"The heavens call to you, and circle around you, displaying to you their eternal splendors."

From *The Divine Comedy*, Dante Aligheri
Italian poet, 1265–1321

Using the Monthly Star Charts

These month-by-month charts of the night sky are the essential outdoor guides for identifying the bright stars and the constellations.

KEY TO SYMBOLS

Magnitudes

Brighter than 0.0

0.0   3.0

       3.5

0.5   4.0

       4.5

1.0   5.0

       5.5

1.5

2.0   6.0

2.5   6.5

Double stars

Variable stars

Open clusters

Globular clusters

Planetary nebulas

Bright nebulas

Dark nebulas

Galaxies

The monthly sky maps chart the entire night sky visible through the year. The charts come in two sets: one for the Northern Hemisphere and one for the Southern.

The Northern Hemisphere charts (pp. 264–87) are for the latitudes of the United States, Canada, Europe, and Japan. While they show the sky from a latitude of 40°N, the charts are usable for latitudes 10 to 15 degrees north or south of this.

The Southern Hemisphere charts (pp. 288–311) depict the sky from a latitude of 35°S. These are for use in the South Pacific, Australia, New Zealand, South America, and southern Africa.

Each month's sky is divided into two halves—showing the sky as you would see it facing north and facing south.

Select the hemisphere you live in, then the current month. Each pair of maps for that month depicts the night sky you will see as you face north or south.

Refer to the description for highlights of what to look for in that month's night sky.

The charts show the sky on these dates and times. (When in force, daylight saving time is shown in italics on the Northern Hemisphere charts; southern observers should adjust where necessary.) The stars visible as the month begins appear two hours earlier by month's end.

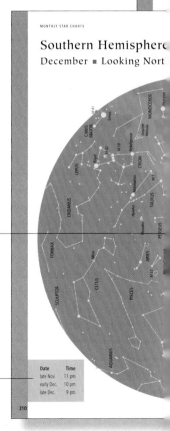

Southern Hemisphere
December ■ Looking North

Date	Time
late Nov.	11 pm
early Dec.	10 pm
late Dec.	9 pm

310

misphere
king North

EAST

the northeastern
thin Ursa Major, the
heast. A line from
d left parallel to the
orth Star. On the

NORTH

URSA MAJOR

LEO
MINOR

CANES
VENATICI

COMA BERENICES

URSA
MINOR

DRACO

CEPHEUS

side of Polaris opposite the Big Dipper lies the bright "W"
of Queen Cassiopeia, while between Cassiopeia and
Polaris hides the faint house-shaped pattern of Cassiopeia's
consort, Cepheus the King. Directly overhead lies Capella,
the brightest star of Auriga the Charioteer.

Northern Hemisphere
February ■ Looking South

Auriga

EAST

SEXTANS

HYDRA

PYXIS

PUPPIS

COLUMBA

CAELIUM

HOROLOGIUM

CANCER

CANIS MINOR

GEMINI

MONOCEROS

CANIS MAJOR

LEPUS

ORION

ERIDANUS

AURIGA

TAURUS

PERSEUS

ARIES

PISCES

SOUTH

More bright stars adorn the sky now than in any other
season. Due south, Orion the Hunter contains Betelgeuse
and Rigel. Orion's three Belt stars form a pointer to other
bright stars: the Belt points down to Sirius in Canis Major,
the Great Dog, the brightest star in the night sky, and

points up toward Ald
Bull. High in the east
brightness—Castor a
Between them and S
Canis Minor, the Litt

Southern Hemisphere
December ■ Looking South

Taurus

EAST

PYXIS

ANTLIA

VELA

CENTAURUS

CANIS MAJOR

PUPPIS

COLUMBA

CAELIUM

CARINA

PICTOR

DORADO

HOROLOGIUM

CHAMAELEON

VOLANS

RETICULUM

MENSA

HYDRUS

MUSCA AUSTRALIS

TRIANGULUM
AUSTRALE

APUS

OCTANS

TUCANA

ERIDANUS

PHOENIX

ARA

PAVO

INDUS

SCULPTOR

TELESCOPIUM

MICROSCOPIUM

GRUS

CORONA
AUSTRALIS

SAGITTARIUS

PISCIS
AUSTRINUS

CAPRICORNUS

SOUTH

The faint constellations of spring make way for the
brighter star patterns of summer now rising into the east.
In the transition, the two Magellanic Clouds (the LMC
and SMC) now appear at their best, high in the south just
above the south celestial pole. From a dark site, these are

rewarding objects for binoculars or a telescope at low
power. Above the SMC shines Achernar, while to the east
of the LMC blazes Canopus, the second brightest star in
the night sky. The lone bright star in the west is Fomalhaut
in Piscis Austrinus, the Southern Fish.

---

down in the Southern Hemisphere sky, with its head and
shoulders below the Belt and his legs above.

and Canis Major, the Great Dog, now rising in the east.
Nearly due south lies an eye-catching cluster of stars, the

Star magnitudes
-1 0 1
2 3 4 5

The dimmest stars
shown on each chart are
magnitude 5, the faintest
visible from locations
near city lights.

Stand facing due north
or south, then hold the
appropriate chart turned
so the horizon lies at the
bottom. Stars directly
overhead lie at the top
of the chart.

Northern Hemisphere
January ■ Looking North

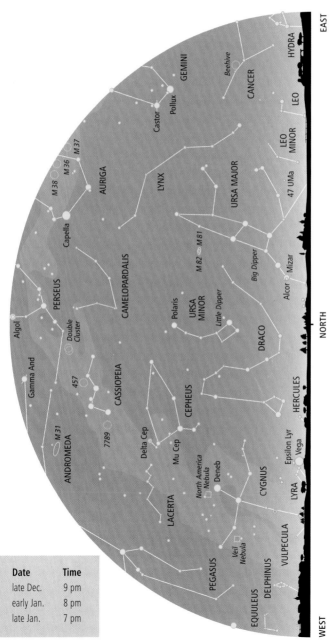

EAST

NORTH

WEST

Date	Time
late Dec.	9 pm
early Jan.	8 pm
late Jan.	7 pm

The Big Dipper, the well-known pattern of stars within Ursa Major, the Great Bear, stands on its handle low on the northeast horizon. Meanwhile, the Little Dipper hangs by its handle from Polaris, the North Star. Cassiopeia the Queen clings to her throne as she hangs upside down high in the north. As you face north, Cassiopeia's familiar "W" of five stars now looks more like a celestial letter M. The twin stars of Castor and Pollux in Gemini the Twins are rising in the eastern sky while Capella, in Auriga the Charioteer, shines down from high in the northeast.

Northern Hemisphere
January ■ Looking South

Orion

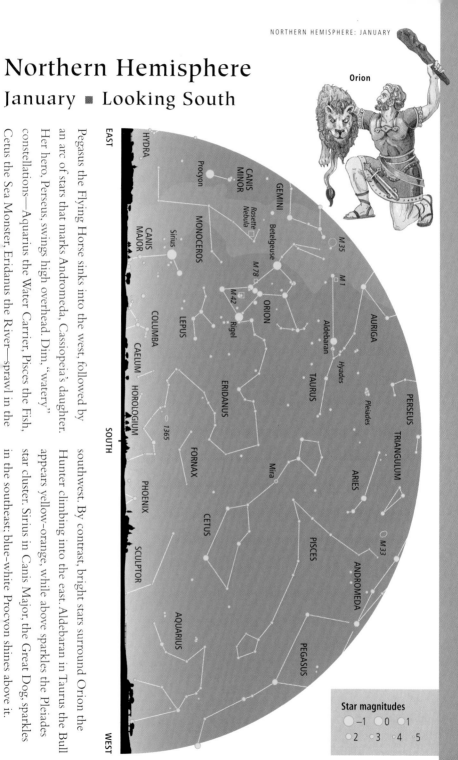

Pegasus the Flying Horse sinks into the west, followed by an arc of stars that marks Andromeda, Cassiopeia's daughter. Her hero, Perseus, swings high overhead. Dim, "watery" constellations—Aquarius the Water Carrier, Pisces the Fish, Cetus the Sea Monster, Eridanus the River—sprawl in the southwest. By contrast, bright stars surround Orion the Hunter climbing into the east. Aldebaran in Taurus the Bull appears yellow-orange, while above sparkles the Pleiades star cluster. Sirius in Canis Major, the Great Dog, sparkles in the southeast; blue-white Procyon shines above it.

EAST

SOUTH

WEST

Star magnitudes
- –1
- 0
- 1
- 2
- 3
- 4
- 5

265

Northern Hemisphere
February ■ Looking North

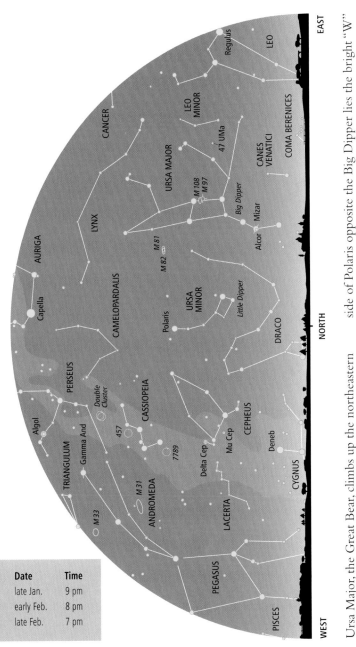

Ursa Major, the Great Bear, climbs up the northeastern sky as he endlessly circles the pole. Within Ursa Major, the Big Dipper rises bowl-first in the northeast. A line from the Dipper's two Pointer stars extended left parallel to the horizon aims straight at Polaris, the North Star. On the side of Polaris opposite the Big Dipper lies the bright "W" of Queen Cassiopeia, while between Cassiopeia and Polaris hides the faint house-shaped pattern of Cassiopeia's consort, Cepheus the King. Directly overhead lies Capella, the brightest star of Auriga the Charioteer.

Date	Time
late Jan.	9 pm
early Feb.	8 pm
late Feb.	7 pm

Northern Hemisphere
February ■ Looking South

More bright stars adorn the sky now than in any other season. Due south, Orion the Hunter contains Betelgeuse and Rigel. Orion's three Belt stars form a pointer to other bright stars: the Belt points down to Sirius in Canis Major, the Great Dog, the brightest star in the night sky, and

points up toward Aldebaran, the reddish eye of Taurus the Bull. High in the eastern sky shine two stars of equal brightness—Castor and Pollux in Gemini the Twins. Between them and Sirius sparkles isolated Procyon, in Canis Minor, the Little Dog.

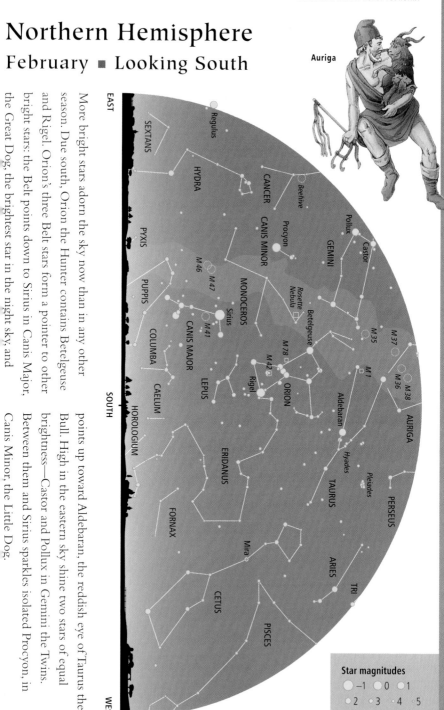

Auriga

EAST

SOUTH

WEST

Regulus

SEXTANS

HYDRA

PYXIS

PUPPIS

CANCER

CANIS MINOR

Procyon

Beehive

GEMINI

Pollux

Castor

M46

M47

MONOCEROS

Rosette Nebula

Sirius

M41

CANIS MAJOR

COLUMBA

CAELUM

HOROLOGIUM

LEPUS

Betelgeuse

M78

M42

ORION

Rigel

M35

M37

M36

M38

M1

AURIGA

Aldebaran

Hyades

TAURUS

Pleiades

PERSEUS

ERIDANUS

FORNAX

Mira

CETUS

PISCES

ARIES

TRI

Star magnitudes

–1 0 1

2 3 4 5

267

Northern Hemisphere
March ■ Looking North

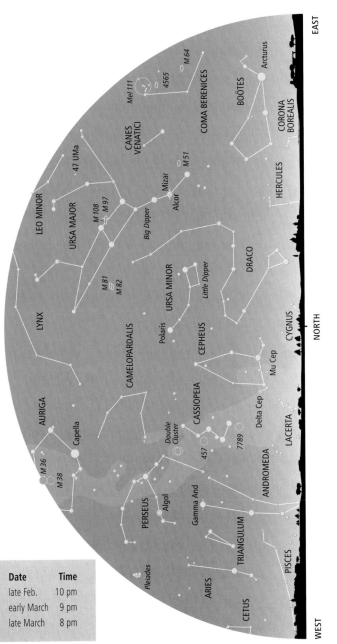

Date	Time
late Feb.	10 pm
early March	9 pm
late March	8 pm

Ursa Major, the Great Bear, now walks upside down high in the north. The brightest seven stars within the Bear form the familiar Big Dipper pattern. The Dipper's two Pointer stars in its Bowl aim down to Polaris, the North Star. Swinging off to the right, or east, of Polaris six other stars form the Little Dipper, also called the Little Bear or Ursa Minor. While well known, the Little Dipper is faint and hard to pick out of urban skies. Only Polaris and the two stars forming the pouring lip of the Little Dipper—the "Guardian" stars—are bright enough to see in the city.

Northern Hemisphere
March ■ Looking South

Ursa Minor

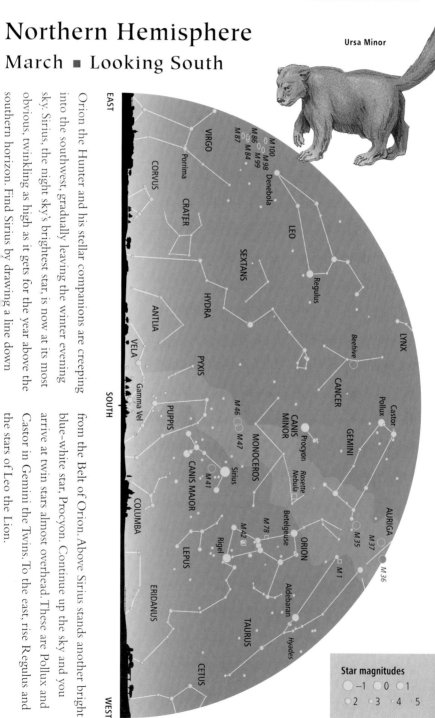

Orion the Hunter and his stellar companions are creeping into the southwest, gradually leaving the winter evening sky. Sirius, the night sky's brightest star, is now at its most obvious, twinkling as high as it gets for the year above the southern horizon. Find Sirius by drawing a line down from the Belt of Orion. Above Sirius stands another bright blue-white star, Procyon. Continue up the sky and you arrive at twin stars almost overhead. These are Pollux and Castor in Gemini the Twins. To the east, rise Regulus and the stars of Leo the Lion.

EAST

SOUTH

WEST

VIRGO
Porrima
CORVUS
CRATER
M 100
M 98
M 86
M 99
M 84
M 87
Denebola
LEO
SEXTANS
Regulus
HYDRA
ANTLIA
VELA
Gamma Vel
PYXIS
PUPPIS
Beehive
CANCER
LYNX
Castor
Pollux
GEMINI
M 46
M 47
CANIS
MINOR
Procyon
MONOCEROS
Rosette
Nebula
Sirius
M 41
CANIS MAJOR
COLUMBA
LEPUS
M 78
Betelgeuse
ORION
Rigel
M 42
AURIGA
M 35
M 37
M 36
M 1
Aldebaran
TAURUS
Hyades
ERIDANUS
CETUS

Star magnitudes
○ −1   ○ 0   ○ 1
○ 2   ○ 3   · 4   · 5

Northern Hemisphere
April ▪ Looking North

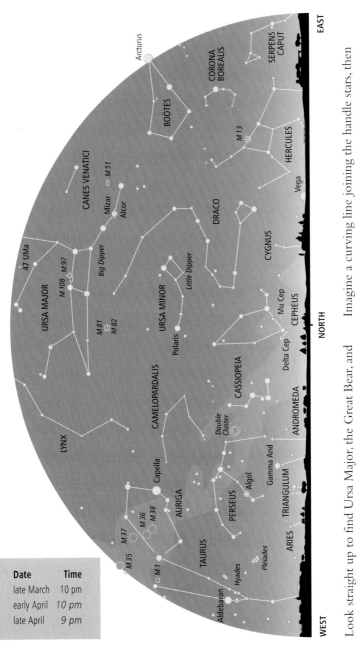

Date	Time
late March	10 pm
early April	*10 pm*
late April	*9 pm*

Look straight up to find Ursa Major, the Great Bear, and the Big Dipper, now upside down. Its two Pointer stars aim straight down to Polaris, the brightest star in Ursa Minor, the Little Bear or Little Dipper. Use the Big Dipper's handle to find the spring sky's brightest star.

Imagine a curving line joining the handle stars, then extend that curve down to Arcturus ("arc to Arcturus"), in Boötes the Herdsman, a kite-shaped pattern rising in the east. As Arcturus rises, look for Capella, the brightest star in Auriga the Charioteer, sinking into the northwest.

Northern Hemisphere
April ■ Looking South

The bright winter stars are all sinking quickly into the western twilight. Replacing them in the south are the less spectacular constellations of spring. High in the south lazes the regal pattern of Leo the Lion and his bright star Regulus. Look for them by drawing a line down from the

two Pointer stars in the pouring lip of the Big Dipper bowl. Find spring's other bright stars by drawing an arc from the Dipper handle to Arcturus, then continuing that arc to Spica, in Virgo the Maiden. To the west of Spica lies the trapezoid-shaped Corvus the Crow.

Boötes

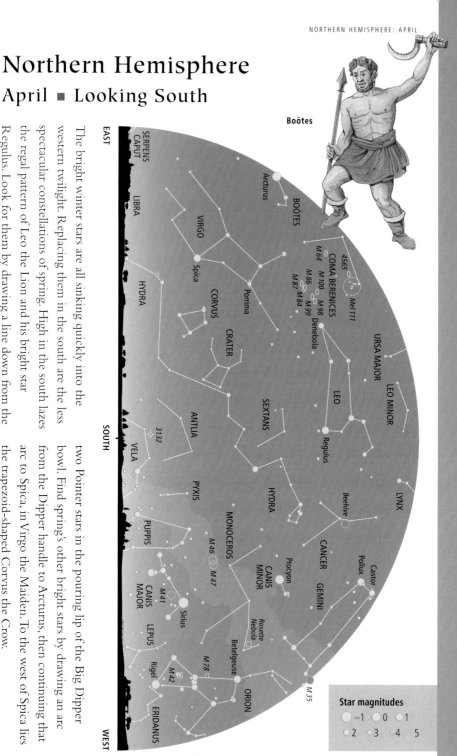

EAST

SERPENS CAPUT

LIBRA

Arcturus

BOÖTES

VIRGO

Spica

Porrima

HYDRA

CORVUS

CRATER

4565
M 64
COMA BERENICES
M 86 M 100 M 98
M 87 M 84 M 99
Denebola
Mel 111

URSA MAJOR

LEO MINOR

LEO

Regulus

LYNX

SEXTANS

ANTLIA

3132

VELA

PYXIS

PUPPIS

HYDRA

MONOCEROS

M 46   M 47

CANCER

Beehive

Procyon

CANIS
MINOR

GEMINI

Castor

Pollux

M 41

CANIS
MAJOR

Sirius

LEPUS

Rigel

M 42

ERIDANUS

Rosette
Nebula

Betelgeuse

M 78

ORION

M 35

SOUTH

WEST

Star magnitudes
●−1 ●0 ●1
●2 ·3 ·4 ·5

271

Northern Hemisphere
May ■ Looking North

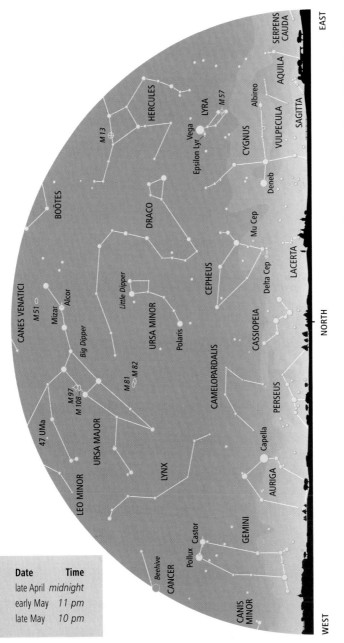

WEST

EAST

NORTH

Date	Time
late April	*midnight*
early May	*11 pm*
late May	*10 pm*

Capella, a bright star of winter, swings close to the north-western horizon in the early evening hours, sparkling like a jewel low in the sky. Also skimming the northern horizon, but due north, lies the stellar "W" of Cassiopeia the Queen. Meanwhile, Deneb and Vega, stars of summer, are beginning to climb into the northeast. The Big Dipper lies upside down high overhead, its two Pointer stars aiming at Polaris and the Little Dipper. From a rural site you may also be able to trace Draco the Dragon's faint stars winding between the Dippers.

Northern Hemisphere
May ■ Looking South

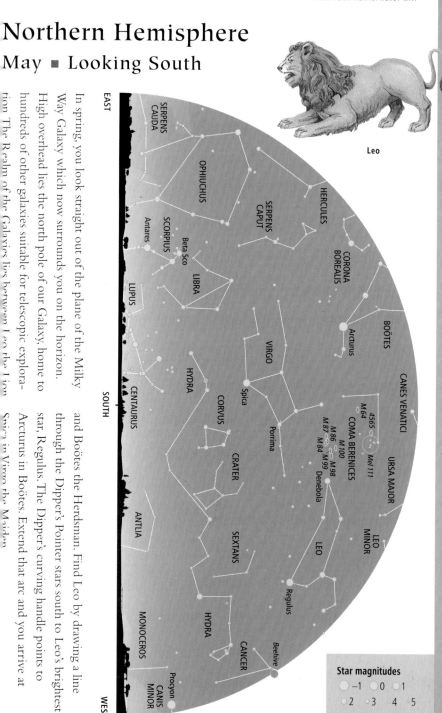

Leo

EAST

SOUTH

WEST

In spring, you look straight out of the plane of the Milky Way Galaxy which now surrounds you on the horizon. High overhead lies the north pole of our Galaxy, home to hundreds of other galaxies suitable for telescopic exploration. The Realm of the Galaxies lies between Leo the Lion

and Boötes the Herdsman. Find Leo by drawing a line through the Dipper's Pointer stars south to Leo's brightest star, Regulus. The Dipper's curving handle points to Arcturus in Boötes. Extend that arc and you arrive at Spica in Virgo the Maiden.

Star magnitudes
○ −1 ○ 0 ○ 1
○ 2 ○ 3 · 4 · 5

Northern Hemisphere
June ■ Looking North

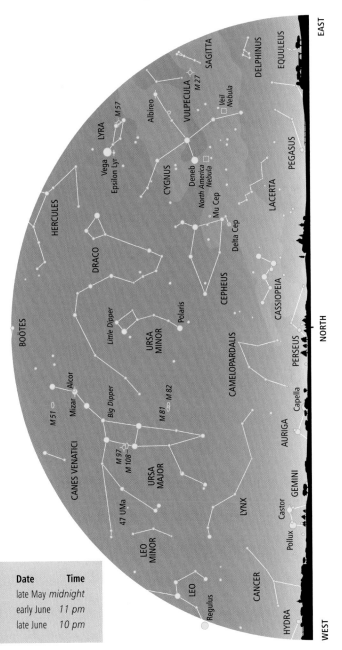

WEST

NORTH

EAST

Date	Time
late May	*midnight*
early June	*11 pm*
late June	*10 pm*

Ursa Major, the Great Bear, and the Big Dipper lie high in the northwest, to the left of Polaris as you face north, while the Little Dipper swings above Polaris. If your skies are dark, between the two Dippers you may be able to trace out a winding chain of stars that marks the wriggling form of Draco the Dragon. His head lies in a box of four stars near Vega, the bright blue-white star now dominating the eastern sky. Below Vega, low in the northeast, look for Deneb in Cygnus the Swan, while the "W" of Cassiopeia the Queen skims low across the northern sky.

Northern Hemisphere
June ■ Looking South

Leo the Lion wanders off into the west carrying Regulus with it. Arcturus and Spica, the brightest stars of spring, are being joined by a new group of stellar companions, the more numerous bright stars of summer. Look for an orange star low in the south. That's Antares, the heart of

Scorpius the Scorpion crawling along the southern horizon. Between Arcturus and Vega, now rising in the east, look for a crooked "H" that marks the main part of Hercules. Between Hercules and Arcturus lies a semicircle of stars, Corona Borealis, the Northern Crown.

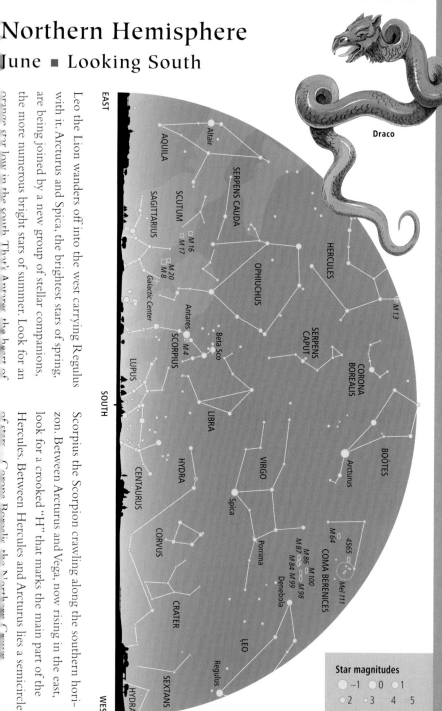

Draco

EAST

SOUTH

WEST

AQUILA
Altair
SERPENS CAUDA
SCUTUM
SAGITTARIUS
M 16
M 17
M 20
M 8
Galactic Center
OPHIUCHUS
HERCULES
M 13
Antares
M 4
SCORPIUS
Beta Sco
SERPENS CAPUT
CORONA BOREALIS
LUPUS
LIBRA
BOÖTES
Arcturus
VIRGO
HYDRA
Spica
CENTAURUS
Porrima
45565
M 64
COMA BERENICES
M 86 M 100
M 87 M 99
M 84 M 98
Mel 111
CORVUS
Denebola
CRATER
LEO
Regulus
SEXTANS
HYDRA

Star magnitudes
–1　0　1
2　3　4　5

275

Northern Hemisphere
July ■ Looking North

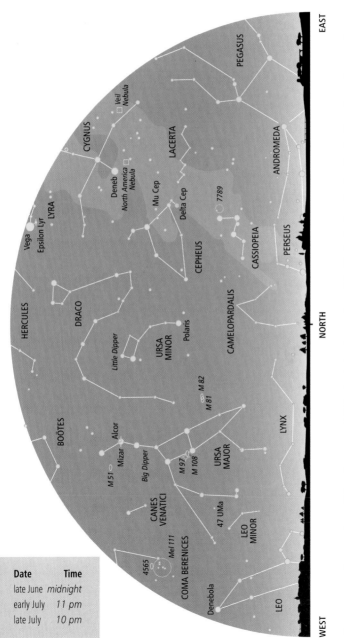

Date	Time
late June	*midnight*
early July	*11 pm*
late July	*10 pm*

In summer, the turning of the sky around Polaris sinks the Big Dipper into the northwest, swings the head of Draco the Dragon overhead, and raises Cassiopeia the Queen into the northeast. In the east, Vega, in Lyra the Harp, and

sky with Altair, in Aquila the Eagle, found on the southern map (see facing page). These three stars, the brightest of summer, make up the Summer Triangle. Vega and Altair lie close by, only 25 and 17 light-years away respectively, but

Northern Hemisphere
July ■ Looking South

Straight overhead look for the crooked "H" pattern of Hercules. Arcturus and the sparse stars of spring still dominate the western sky. Due south, look for Antares amid the hook-shaped Scorpius the Scorpion. To the left,

this direction you are looking toward the center of our Galaxy. If the sky is dark (with no Moon) and haze-free, you will see what looks like steam rising from the teapot and wafting up the sky past the stars Altair and Vega. That's

Lyra

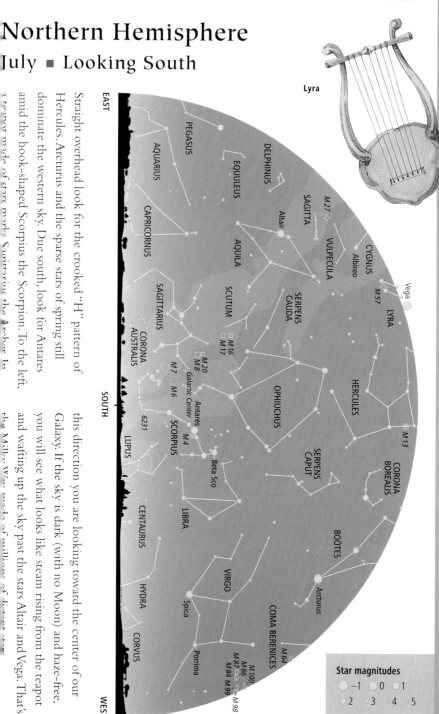

EAST

SOUTH

WEST

PEGASUS

AQUARIUS

CAPRICORNUS

EQUULEUS

DELPHINUS

SAGITTA

Altair

VULPECULA

AQUILA

SCUTUM

M 27

Albireo

CYGNUS

SERPENS
CAUDA

Vega

M 57

LYRA

SAGITTARIUS

CORONA
AUSTRALIS

M 16
M 17

M 20
M 8
Galactic Center

HERCULES

M 7   M 6

6231

LUPUS

Antares
M 4

SCORPIUS

Beta Sco

OPHIUCHUS

SERPENS
CAPUT

M 13

CORONA
BOREALIS

CENTAURUS

LIBRA

BOÖTES

HYDRA

VIRGO

Spica

Porrima

COMA BERENICES

Arcturus

M 64
M 100
M 86
M 87   M 99
M 84   M 98

CORVUS

Star magnitudes
–1 0 1
2 3 4 5

Northern Hemisphere
August ■ Looking North

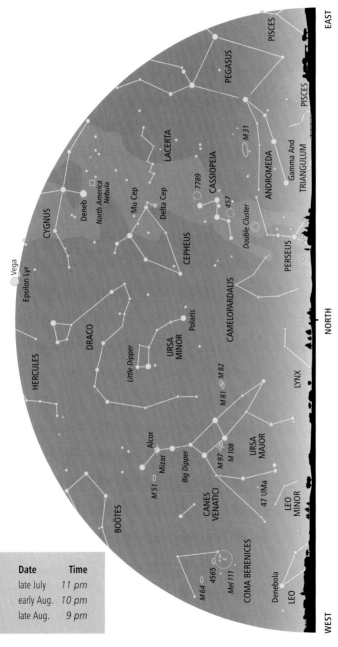

WEST

NORTH

EAST

Vega
Epsilon Lyr

HERCULES

CYGNUS
Deneb
North America Nebula

DRACO

CEPHEUS
Mu Cep
Delta Cep

LACERTA

PEGASUS

PISCES

7789
CASSIOPEIA
457

PISCES

Little Dipper

URSA MINOR
Polaris

CAMELOPARDALIS

Double Cluster
PERSEUS

ANDROMEDA
M 31

Gamma And
TRIANGULUM

M 81 — *M 82*

LYNX

Alcor
Mizar
M 51
Big Dipper
M 97
M 108
URSA MAJOR
47 UMa

BOÖTES

CANES VENATICI

LEO MINOR

M 64
4565
Mel 111
COMA BERENICES

Denebola
LEO

Date	Time
late July	11 pm
early Aug.	10 pm
late Aug.	9 pm

The Big Dipper and Cassiopeia the Queen now lie on a horizontal line, seemingly balanced on opposite sides of the north celestial pole, marked by Polaris, the North Star. Through the night, as Earth spins, the entire sky appears to turn about this. ... Overhead, you will see blue-white

Vega, the brightest star of summer. Together with Deneb, in the east, and Altair, in the south, the three stars form the hallmark of the summer sky, the Summer Triangle. The Milky Way, formed by our Galaxy's spiral arms, rises out of the northeast to sweep through the Triangle.

Northern Hemisphere
August ■ Looking South

Cygnus

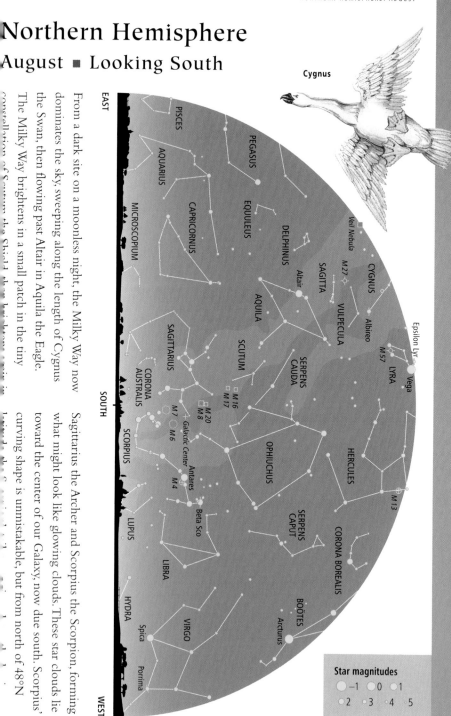

From a dark site on a moonless night, the Milky Way now dominates the sky, sweeping along the length of Cygnus the Swan, then flowing past Altair in Aquila the Eagle. The Milky Way brightens in a small patch in the tiny constellation of Scutum the Shield, then brightens again in

Sagittarius the Archer and Scorpius the Scorpion, forming what might look like glowing clouds. These star clouds lie toward the center of our Galaxy, now due south. Scorpius' curving shape is unmistakable, but from north of 48°N

EAST

SOUTH

WEST

PISCES
AQUARIUS
PEGASUS
MICROSCOPIUM
CAPRICORNUS
EQUULEUS
DELPHINUS
Altair
SAGITTA
AQUILA
SCUTUM
SAGITTARIUS
CORONA
AUSTRALIS
Veil Nebula
M 27
CYGNUS
VULPECULA
Albireo
SERPENS
CAUDA
M 16
M 17
M 20
M 8
M 7
M 6
Galactic Center
SCORPIUS
Antares
M 4
Beta Sco
OPHIUCHUS
LUPUS
LIBRA
HYDRA
Epsilon Lyr
Vega
M 57
LYRA
HERCULES
M 13
SERPENS
CAPUT
CORONA BOREALIS
BOÖTES
Arcturus
VIRGO
Spica
Porrima

Star magnitudes
−1 0 1
2 3 4 5

Northern Hemisphere
September ■ Looking North

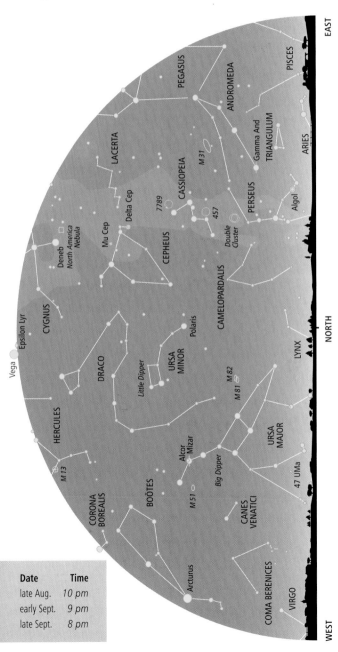

EAST

PEGASUS

LACERTA

ANDROMEDA

CASSIOPEIA

M31

Gamma And

TRIANGULUM

7789

PERSEUS

Delta Cep

PISCES

457

Mu Cep

CEPHEUS

Double Cluster

Algol

ARIES

Deneb
North America Nebula

CYGNUS

CAMELOPARDALIS

Epsilon Lyr

Vega

DRACO

Polaris

Little Dipper

URSA MINOR

LYNX

NORTH

HERCULES

M 82

M 81

M 13

Alcor
Mizar

URSA MAJOR

M 51

Big Dipper

47 UMa

CORONA BOREALIS

BOÖTES

CANES VENATICI

Arcturus

COMA BERENICES

VIRGO

WEST

Date	Time
late Aug.	10 pm
early Sept.	9 pm
late Sept.	8 pm

Ursa Major, the Great Bear, now appears to be lumbering along the northern horizon. The Big Dipper, contained within Ursa Major, swings below the pole while the Little Dipper and the head of Draco the Dragon swing above

Cassiopeia the Queen, Andromeda the Chained Princess, and the Square of Pegasus the Flying Horse—are beginning to rise into the evening sky, gradually pushing the summer stars off stage. The Milky Way now bisects the

Northern Hemisphere
September ■ Looking South

The Summer Triangle stars, Vega, Altair, and to the north, Deneb, sparkle high overhead in the evening hours. Vega belongs to Lyra the Harp, a small parallelogram of stars. Altair belongs to Aquila the Eagle, depicted flying north along the Milky Way. To the right, or west, of Vega look

for the H-shaped pattern of Hercules. To the west of Altair, stands Ophiuchus the Serpent Bearer. Below Altair, on the southern horizon between Sagittarius the Archer and Scorpius the Scorpion, lies the glowing heart of the

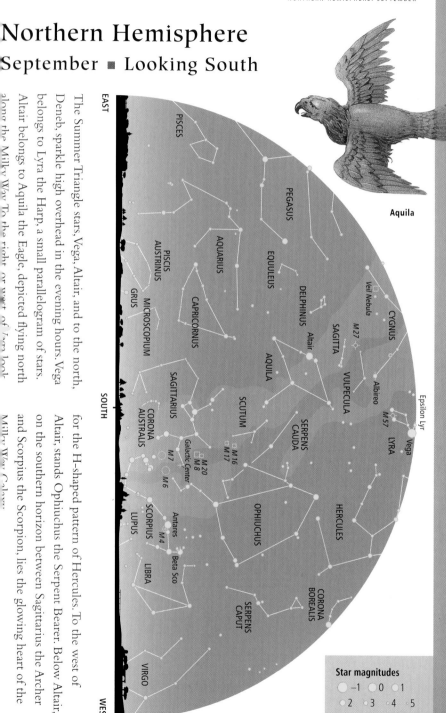

Aquila

EAST

SOUTH

WEST

PISCES

PEGASUS

PISCES
AUSTRINUS

AQUARIUS

EQUULEUS

GRUS

MICROSCOPIUM

CAPRICORNUS

DELPHINUS

Veil Nebula

M 27

SAGITTA

CYGNUS

Albireo

M 57

LYRA

Vega

Epsilon Lyr

Altair

AQUILA

VULPECULA

SAGITTARIUS

SCUTUM

SERPENS
CAUDA

CORONA
AUSTRALIS

M 16
M 17
Galactic Center
M 20
M 8

M 7
M 6

HERCULES

OPHIUCHUS

SERPENS
CAPUT

CORONA
BOREALIS

SCORPIUS

Antares

M 4

Beta Sco

LUPUS

LIBRA

VIRGO

Star magnitudes
−1 0 1
2 3 4 5

Northern Hemisphere
October ■ Looking North

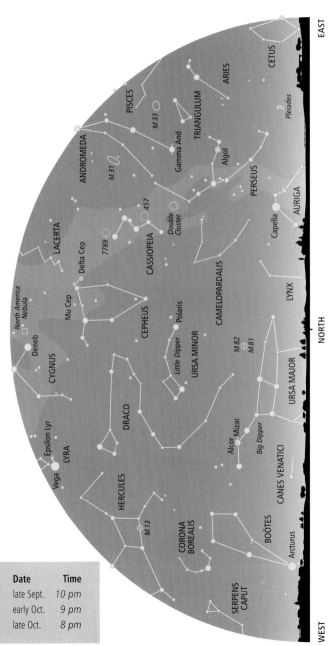

Date	Time
late Sept.	10 pm
early Oct.	9 pm
late Oct.	8 pm

The Big Dipper skims the horizon, while opposite the pole, Cassiopeia the Queen climbs high into the northeast. Below the "W" of Cassiopeia shine the stars of Andromeda, her daughter, and Perseus, the hero who rescued Andromeda from Cetus the Sea Monster

itself visible low in the south. Deneb, the most northerly star in the Summer Triangle, shines down from directly overhead. Two bright stars twinkle on the northern horizon: Arcturus setting in the northwest, and Capella rising in the northeast.

Northern Hemisphere
October ■ Looking South

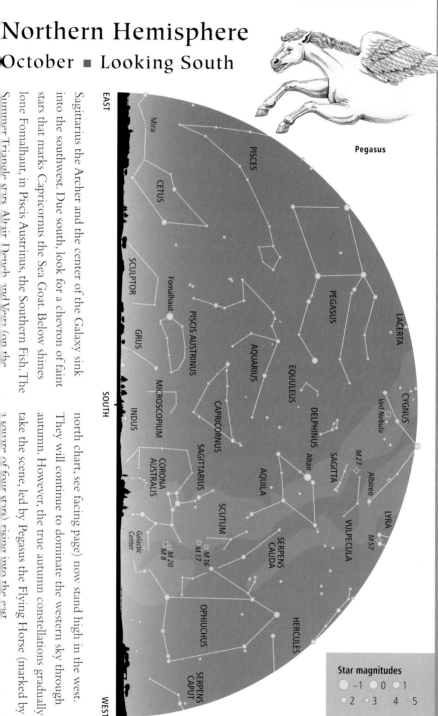

Pegasus

EAST

SOUTH

WEST

Mira

PISCES

CETUS

SCULPTOR

Fomalhaut

GRUS

PISCIS AUSTRINUS

MICROSCOPIUM

INDUS

AQUARIUS

CAPRICORNUS

CORONA AUSTRALIS

SAGITTARIUS

SCUTUM

M 20
M 8

M 16
M 17

Galactic Center

OPHIUCHUS

SERPENS CAPUT

SERPENS CAUDA

AQUILA

EQUULEUS

DELPHINUS

Altair

SAGITTA

PEGASUS

LACERTA

Veil Nebula

CYGNUS

M 27

Albireo

LYRA

M 57

VULPECULA

HERCULES

Sagittarius the Archer and the center of the Galaxy sink into the southwest. Due south, look for a chevron of faint stars that marks Capricornus the Sea Goat. Below shines lone Fomalhaut, in Piscis Austrinus, the Southern Fish. The Summer Triangle stars, Altair, Deneb, and Vega (on the

north chart, see facing page) now stand high in the west. They will continue to dominate the western sky through autumn. However, the true autumn constellations gradually take the scene, led by Pegasus the Flying Horse (marked by a square of four stars) rising into the east.

283

Northern Hemisphere
November ■ Looking North

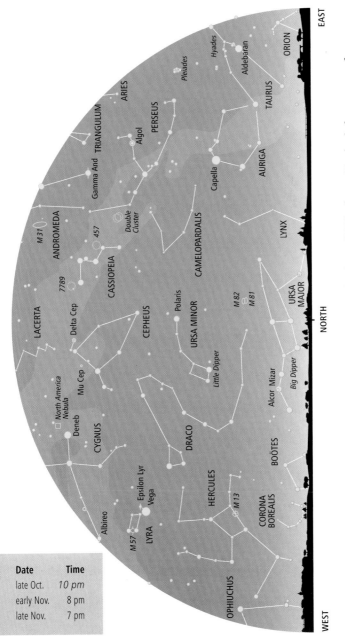

EAST

NORTH

WEST

Date	Time
late Oct.	10 pm
early Nov.	8 pm
late Nov.	7 pm

The Big Dipper is now at its lowest point in the north, so low it may be hiding behind buildings and trees or, for those living south of 35°N latitude, below the horizon. High in the north, however, the five stars of Cassiopeia the Queen are easy to spot. Look for a letter "M" or an

upside-down "W" of stars. To the left, or west, of Cassiopeia lies the faint house-shaped Cepheus the King. To the east of Cassiopeia, a Y-shaped chain of stars marks Perseus the Hero. Below, one of the first stars of winter, Capella, in Auriga the Charioteer, rises into the northeast

Northern Hemisphere
November ▪ Looking South

Perseus

High in the south, look for the square-shaped pattern marking Pegasus the Flying Horse. Connected to the top left corner of the Square, two arcs of stars stretch away to the east. They mark Andromeda the Chained Princess,

Monster, a faint pattern low in the south, as Perseus flew in on the back of Pegasus. Between Cetus and Pegasus shines another faint, "watery" constellation, Pisces the Fish. Another fish, Piscis Austrinus, is marked by a lonely bright

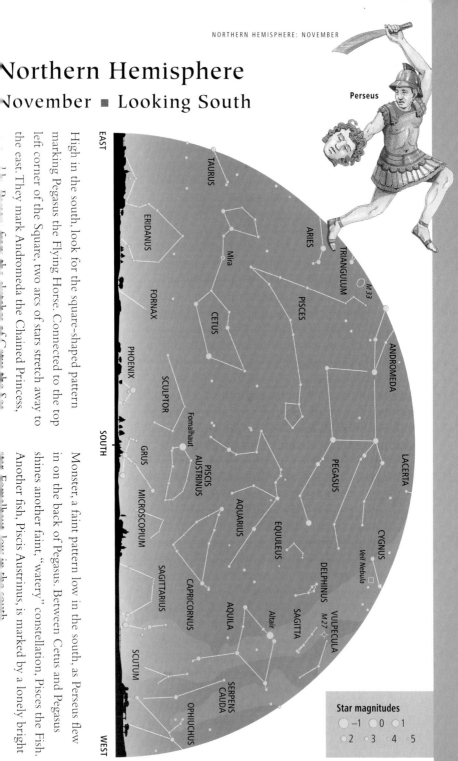

EAST

SOUTH

WEST

TAURUS

ERIDANUS

ARIES

Mira

TRIANGULUM

M 33

PISCES

CETUS

ANDROMEDA

FORNAX

PHOENIX

SCULPTOR

Fomalhaut

PISCIS
AUSTRINUS

GRUS

MICROSCOPIUM

AQUARIUS

PEGASUS

LACERTA

EQUULEUS

CYGNUS

Veil Nebula

DELPHINUS

VULPECULA
M 27

SAGITTARIUS

CAPRICORNUS

AQUILA

Altair

SAGITTA

SCUTUM

SERPENS
CAUDA

OPHIUCHUS

Star magnitudes
○ −1   ○ 0   ○ 1
○ 2   ○ 3   · 4   · 5

Northern Hemisphere
December ■ Looking North

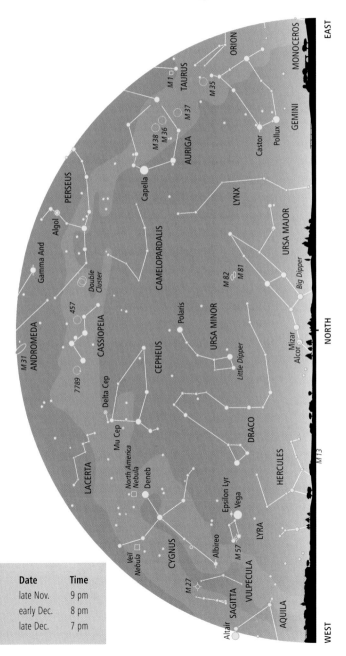

EAST

NORTH

WEST

ORION
TAURUS
M 1
M 35
AURIGA
M 37
M 38
M 36
Capella
GEMINI
Castor
Pollux
MONOCEROS
PERSEUS
Algol
LYNX
Gamma And
CAMELOPARDALIS
Double Cluster
457
URSA MAJOR
ANDROMEDA
M 31
7789
CASSIOPEIA
Polaris
M 82
M 81
Big Dipper
Delta Cep
URSA MINOR
Mizar
Alcor
CEPHEUS
Little Dipper
LACERTA
Mu Cep
DRACO
North America Nebula
Deneb
HERCULES
Epsilon Lyr
Vega
CYGNUS
Albireo
LYRA
Veil Nebula
M 57
M 13
VULPECULA
M 27
SAGITTA
Altair
AQUILA

Date	Time
late Nov.	9 pm
early Dec.	8 pm
late Dec.	7 pm

The three persistent stars of the Summer Triangle, Vega, Altair, and Deneb, can still be seen, although they are now setting into the northwestern sky. The Big Dipper and Ursa Major, the Great Bear, graze the northern horizon while Cassiopeia the Queen now stands as high in the

north as she gets for the year. Stretching to the right, or east, of Cassiopeia, look for the Y-shaped pattern of Perseus the Hero. Below Perseus sparkle some stars of winter: Capella high in the northeast, and the twin stars of Castor and Pollux just rising.

Northern Hemisphere
December ■ Looking South

Pegasus the Flying Horse flies high in the south while Andromeda the Chained Princess reclines overhead. Farther south, the sky is populated by faint, "watery" constellations—Pisces the Fish, Aquarius the Water Bearer, Cetus the Sea Monster, Eridanus the River—all difficult to pick out in bright suburban skies. However, the luminous stars of winter are set to replace the dim stars of autumn. To the east, the star-filled Orion the Hunter and Taurus the Bull, locked in battle, are scuffling into the sky. In Taurus, the star cluster called the Pleiades is sure to catch the eye.

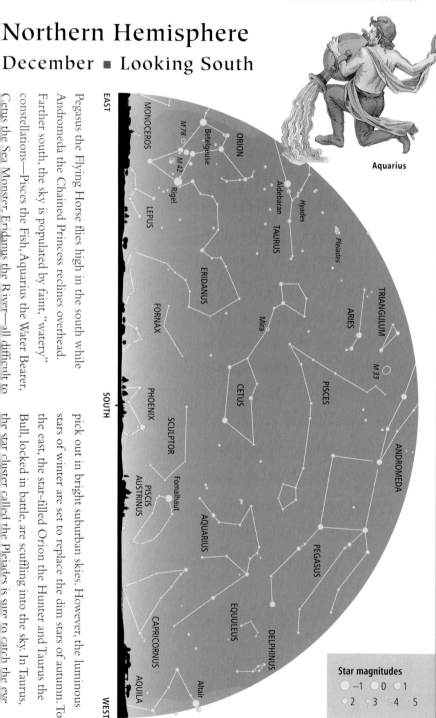

Aquarius

EAST

MONOCEROS

M 78

Betelgeuse

ORION

M 42

Rigel

LEPUS

Aldebaran

Hyades

TAURUS

Pleiades

ERIDANUS

FORNAX

Mira

TRIANGULUM

ARIES

PISCES

M 33

CETUS

ANDROMEDA

PHOENIX

SCULPTOR

Fomalhaut

PISCIS
AUSTRINUS

AQUARIUS

PEGASUS

SOUTH

EQUULEUS

DELPHINUS

CAPRICORNUS

Altair

AQUILA

WEST

Star magnitudes

● −1 ● 0 ● 1
● 2 ● 3 · 4 · 5

287

Southern Hemisphere
January ■ Looking North

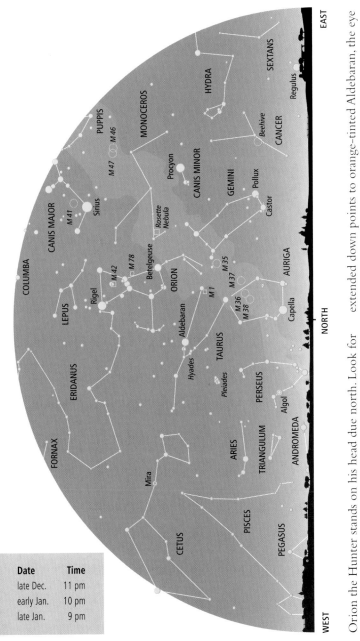

EAST

NORTH

WEST

Date	Time
late Dec.	11 pm
early Jan.	10 pm
late Jan.	9 pm

Orion the Hunter stands on his head due north. Look for Y-shaped Lepus the Hare cowering at his feet. A line drawn through the three stars of Orion's Belt and extended up to the right points to Sirius, the "dog star" and the brightest in the night sky. That same Belt line extended down points to orange-tinted Aldebaran, the eye of Taurus the Bull. Below Aldebaran sparkles a tight cluster of stars, the Pleiades. Capella twinkles just above the northern horizon, with the twin stars of Castor and Pollux rising to the right, or east.

Southern Hemisphere
January ■ Looking South

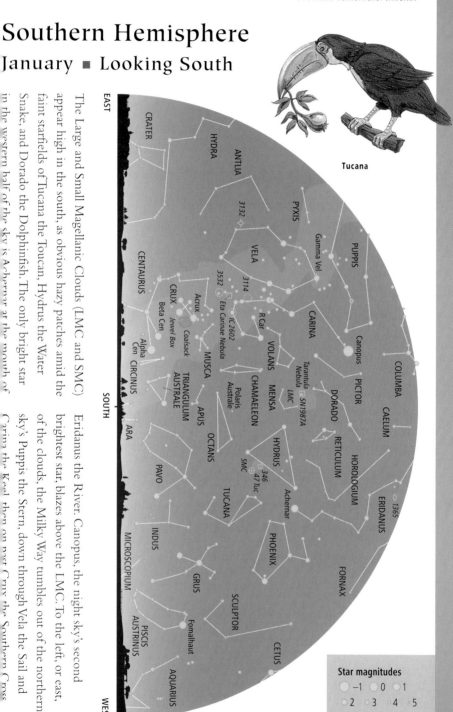

Tucana

The Large and Small Magellanic Clouds (LMC and SMC) appear high in the south, as obvious hazy patches amid the faint starfields of Tucana the Toucan, Hydrus the Water Snake, and Dorado the Dolphinfish. The only bright star in the western half of the sky is Achernar at the mouth of

Eridanus the River. Canopus, the night sky's second brightest star, blazes above the LMC. To the left, or east, of the clouds, the Milky Way tumbles out of the northern sky's Puppis the Stern, down through Vela the Sail and Carina the Keel, then on past Crux the Southern Cross

EAST

CRATER

HYDRA

ANTLIA

3132

PYXIS

Gamma Vel

VELA

3114

3532

CENTAURUS

CRUX

Actrux

Beta Cen

Jewel Box

Coalsack

Alpha
Cen

CIRCINUS

ARA

SOUTH

PAVO

INDUS

MICROSCOPIUM

WEST

R Car

Eta Carinae Nebula

IC 2602

CARINA

VOLANS

MUSCA

TRIANGULUM
AUSTRALE

APUS

OCTANS

CHAMAELEON

MENSA

Polaris
Australe

LMC

Tarantula
Nebula

SN1987A

HYDRUS

SMC

346

47 Tuc

TUCANA

PHOENIX

GRUS

SCULPTOR

PISCIS
AUSTRINUS

AQUARIUS

CETUS

Fomalhaut

PUPPIS

Canopus

PICTOR

DORADO

RETICULUM

COLUMBA

CAELUM

HOROLOGIUM

ERIDANUS

Achernar

1365

FORNAX

Southern Hemisphere
February ■ Looking North

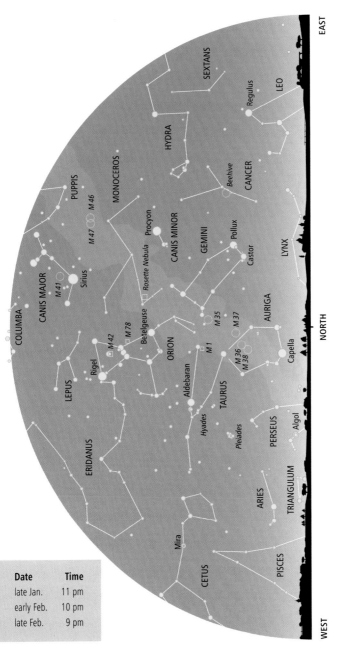

EAST

SEXTANS

Regulus

LEO

HYDRA

Beehive

PUPPIS

CANCER

MONOCEROS

M46

M47

Procyon

CANIS MINOR

Rosette Nebula

M41

Sirius

GEMINI

Pollux

CANIS MAJOR

Castor

LYNX

COLUMBA

M78

Betelgeuse

M35

M37

AURIGA

LEPUS

M42

M1

M36

Rigel

ORION

M38

Capella

Aldebaran

NORTH

ERIDANUS

TAURUS

Hyades

Pleiades

PERSEUS

Algol

ARIES

TRIANGULUM

Mira

CETUS

PISCES

WEST

Date	Time
late Jan.	11 pm
early Feb.	10 pm
late Feb.	9 pm

The northern summer sky is filled with bright stars. Orion the Hunter, and its luminaries Betelgeuse and Rigel, shine high in the north. Above Orion sparkles Sirius in Canis Major, the Great Dog. A lone bright star below Sirius, called Procyon marks Canis Minor the Little Dog. Below Procyon shine Castor and Pollux, the twin stars of Gemini the Twins. Below and to the left, or west, of Orion look for reddish-orange Aldebaran in Taurus the Bull. Capella in Auriga the Charioteer, twinkling just above the northern horizon, rounds out the summer stars

Southern Hemisphere
February ■ Looking South

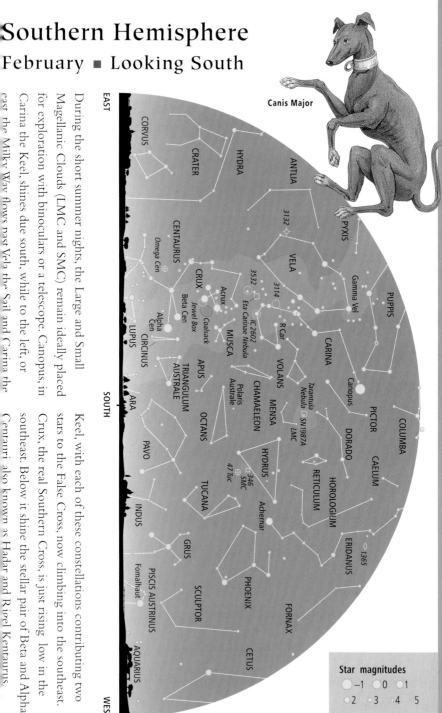

During the short summer nights, the Large and Small Magellanic Clouds (LMC and SMC) remain ideally placed for exploration with binoculars or a telescope. Canopus, in Carina the Keel, shines due south, while to the left, or southeast, the Milky Way flows past Vela the Sail and Carina the

Keel, with each of these constellations contributing two stars to the False Cross, now climbing into the southeast. Crux, the real Southern Cross, is just rising low in the southeast. Below it shine the stellar pair of Beta and Alpha Centauri, also known as Hadar and Rigel Kentaurus.

Canis Major

EAST

SOUTH

WEST

CORVUS
CRATER
HYDRA
ANTLIA
PYXIS
VELA
CENTAURUS
3132
Omega Cen
CRUX
Acrux
3532
3114
Eta Carinae Nebula
IC 2602
R Car
CARINA
Gamma Vel
PUPPIS
Alpha Cen
Beta Cen
Jewel Box
Coalsack
MUSCA
VOLANS
Canopus
PICTOR
COLUMBA
CAELUM
LUPUS
CIRCINUS
APUS
TRIANGULUM AUSTRALE
Polaris Australe
CHAMAELEON
MENSA
Tarantula Nebula
SN1987A
LMC
DORADO
HOROLOGIUM
RETICULUM
ARA
OCTANS
HYDRUS
346
SMC
47 Tuc
Achernar
ERIDANUS
1365
PAVO
TUCANA
INDUS
GRUS
Fomalhaut
PISCIS AUSTRINUS
SCULPTOR
PHOENIX
FORNAX
CETUS
AQUARIUS

Star magnitudes
–1 0 1
2 3 4 5

Southern Hemisphere
March ■ Looking North

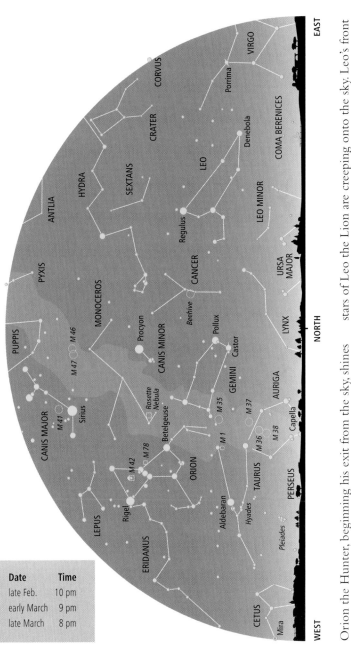

WEST

EAST

NORTH

Date	Time
late Feb.	10 pm
early March	9 pm
late March	8 pm

Orion the Hunter, beginning his exit from the sky, shines in the northwest. Above the trio of stars that form his Belt look for the night sky's brightest star, Sirius, high in the north. Below Sirius shine Procyon and the close pair of Castor and Pollux. Low in the northeast, Regulus and the stars of Leo the Lion are creeping onto the sky. Leo's front is marked by a dangling sickle or hook of stars, while his hindquarters are defined by a right-angle triangle of stars. Between Leo and Gemini the Twins look for a naked-eye star cluster, the Beehive in Cancer the Crab,

Southern Hemisphere
March ■ Looking South

Gemini

The Milky Way is magnificent in autumn. It brightens as it comes out of the north and the constellation of Puppis the Stern now overhead above brilliant Canopus. As it flows past Carina the Keel and Crux the Southern Cross, the Milky Way becomes filled with bright star clouds, clusters,

and nebulas that provide stunning panoramas for binoculars. Below Carina and Crux shine Beta and Alpha Centauri. Nowhere else do we find two stars as bright as these so close together. Beta is 526 light-years away, while Alpha lies a bit more than 4 light-years from Earth.

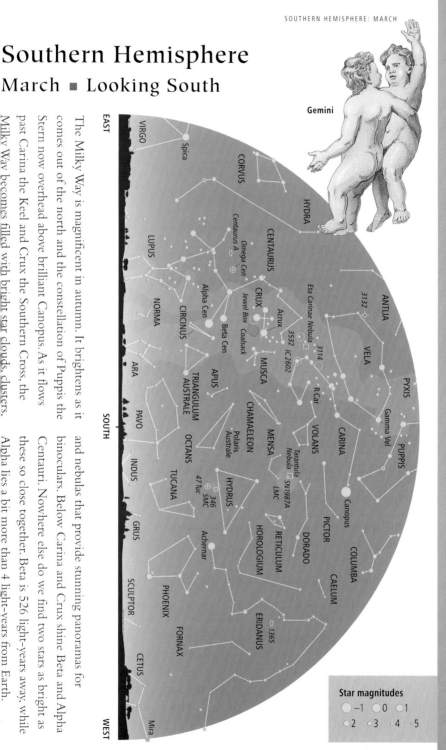

EAST

SOUTH

WEST

VIRGO

Spica

CORVUS

HYDRA

CENTAURUS

Centaurus A

Omega Cen

Jewel Box

CRUX

Acrux

3532

IC 2602

Coalsack

Eta Carinae Nebula

3114

3132

ANTLIA

VELA

PYXIS

Gamma Vel

PUPPIS

LUPUS

Alpha Cen

Beta Cen

MUSCA

R Car

VOLANS

CARINA

Canopus

COLUMBA

NORMA

CIRCINUS

ARA

APUS

TRIANGULUM AUSTRALE

CHAMAELEON

Polaris Australe

MENSA

Tarantula Nebula

LMC

SN1987A

HYDRUS

DORADO

PICTOR

CAELUM

PAVO

OCTANS

TUCANA

47 Tuc

346 SMC

RETICULUM

HOROLOGIUM

ERIDANUS

INDUS

GRUS

Achernar

1365

SCULPTOR

PHOENIX

FORNAX

CETUS

Mira

Star magnitudes
−1 0 1
2 3 4 5

293

Southern Hemisphere
April ■ Looking North

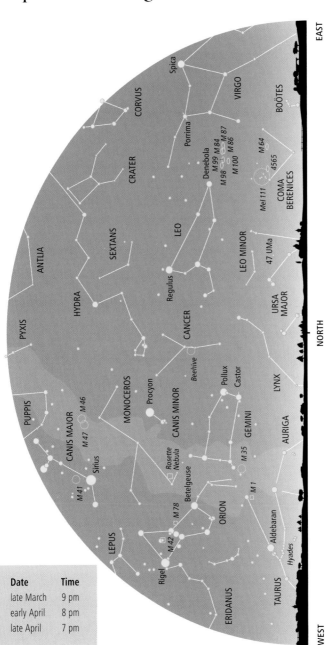

EAST

NORTH

WEST

Date	Time
late March	9 pm
early April	8 pm
late April	7 pm

Orion the Hunter, and the other bright stars of the northern summer sky, such as Sirius and Procyon, are gradually sinking into the western twilight. Taking their place is a relatively sparse area of sky. The most distinctive constellation of the northern autumn sky is Leo the Lion.

The brightest star in the northeast is Regulus, the shining heart of Leo. A hook of stars hanging down from Regulus marks Leo's head and mane (like many older constellations invented in the Northern Hemisphere, Leo appears upside down when viewed from southern latitudes).

Southern Hemisphere
April ▪ Looking South

The four stars of the False Cross stand due south. Though distinctive, it is not a proper constellation. The False Cross's two top stars belong to Vela the Sail, while its bottom two stars belong to Carina the Keel. Canopus, the brilliant star far to the right, or west, also belongs to Carina. Carina,

Vela, Puppis the Stern, and Pyxis the Compass used to make up one huge constellation, the ship Argo Navis. Crux, the real Southern Cross, its stars bright enough to be seen even in city skies, is unmistakable high in the southeast, as are Beta and Alpha Centauri below it.

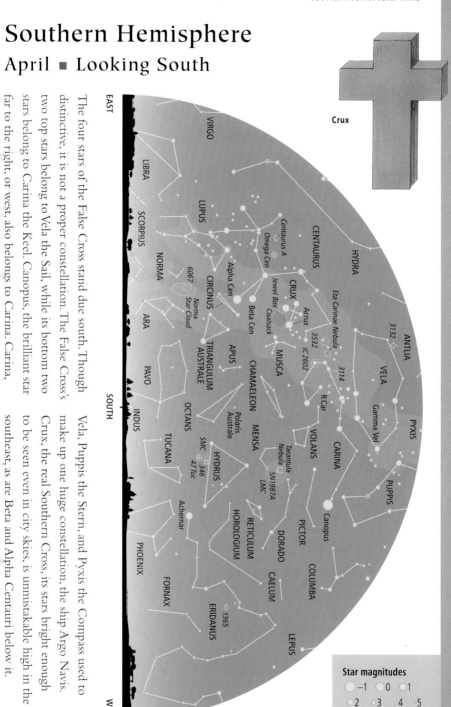

Crux

EAST

SOUTH

WEST

VIRGO
LIBRA
SCORPIUS
LUPUS
NORMA
6067
CIRCINUS
ARA
PAVO
INDUS
TUCANA
OCTANS
TRIANGULUM AUSTRALE
APUS
CHAMAELEON
CENTAURUS
Centaurus A
Omega Cen
Alpha Cen
Beta Cen
Norma Star Cloud
CRUX
Actrux
3532
Jewel Box
Coalsack
IC 2602
3114
MUSCA
Polaris Australe
MENSA
HYDRUS
SMC
346
47 Tuc
Achernar
PHOENIX
FORNAX
1365
ERIDANUS
HOROLOGIUM
RETICULUM
DORADO
CAELUM
COLUMBA
LEPUS
PICTOR
VOLANS
CARINA
Canopus
PUPPIS
PYXIS
Gamma Vel
VELA
3132
ANTLIA
HYDRA
Eta Carinae Nebula
R Car
Tarantula Nebula
SN1987A
LMC

PHOENIX

Star magnitudes
-1 0 1
2 3 4 5

Southern Hemisphere
May ■ Looking North

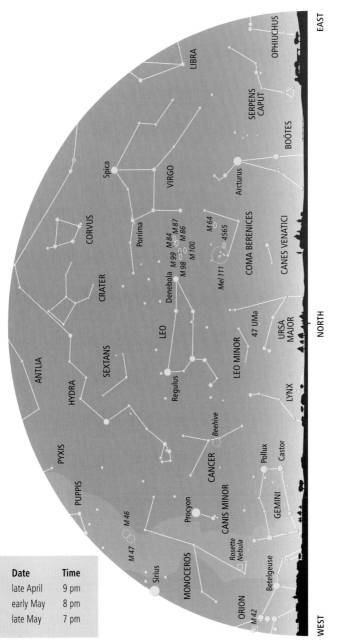

WEST

EAST

NORTH

Date	Time
late April	9 pm
early May	8 pm
late May	7 pm

As you gaze north, you are looking out of the plane of our Milky Way Galaxy (now behind you to the south) toward its north pole. The North Galactic Pole lies near the large naked-eye star cluster Melotte 111, also called the Coma Berenices Cluster. In this direction, you are looking through the thinnest part of our Galaxy's disk. So, looking north we see a sky with few bright stars. Regulus and the stars of Leo the Lion stand due north, while to the right, or east, shine Arcturus low in the sky in Boötes the Herdsman, and Spica high in the east in Virgo the Maiden.

Southern Hemisphere
May ■ Looking South

There is no finer stellar panorama than what now lies to the south. The Milky Way and the plane of our Galaxy pours across the sky, from Sirius in the west to Antares rising in the east. Crux the Southern Cross stands upright due south, its long axis pointing down to the south

celestial pole in Octans the Octant. To the east shine Alpha and Beta Centauri, also called the Pointers because a line drawn through them points to the Cross. The rest of Centaurus the Centaur, sprawling above and around Crux, prances high in the southeast.

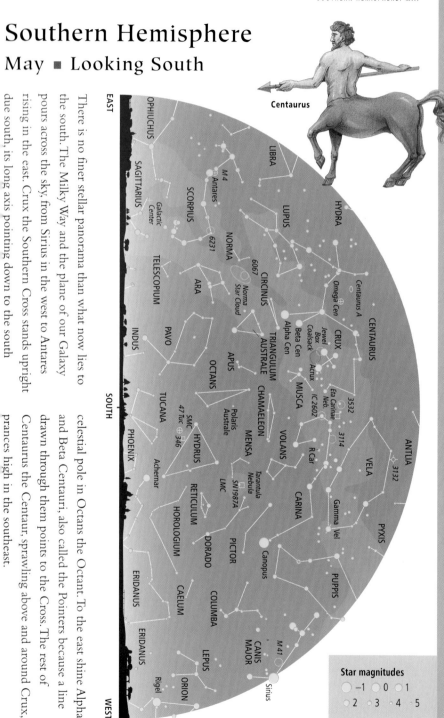

Centaurus

EAST

OPHIUCHUS

SAGITTARIUS

LIBRA

M4
Antares
Galactic
Center
SCORPIUS

LUPUS

HYDRA

Centaurus A

6231

NORMA

TELESCOPIUM

6067

CIRCINUS
Norma
Star Cloud

ARA

INDUS

PAVO

TRIANGULUM
AUSTRALE
Alpha Cen

Beta Cen
Coalsack
Acrux

Omega Cen

CENTAURUS
Jewel
Box
CRUX

Eta Carinae
Neb.
IC 2602

3532

3114

APUS

OCTANS

CHAMAELEON

MUSCA

VOLANS

R Car

SOUTH

TUCANA

SMC
47 Tuc
346

Polaris
Australe

MENSA

CARINA

ANTLIA

3132

VELA

Gamma Vel

PHOENIX

HYDRUS

Achernar

Tarantula
Nebula
SN 1987A
LMC

RETICULUM

HOROLOGIUM

DORADO

PICTOR

Canopus

PYXIS

PUPPIS

ERIDANUS

CAELUM

COLUMBA

CANIS
MAJOR

M41

Sirius

ERIDANUS

Rigel

LEPUS

ORION

WEST

Star magnitudes

–1 0 1
2 3 4 5

Southern Hemisphere

June ■ Looking North

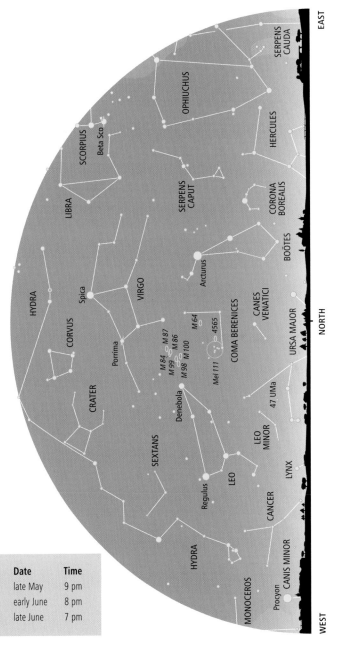

EAST

NORTH

WEST

Date	Time
late May	9 pm
early June	8 pm
late June	7 pm

Two bright stars dominate the otherwise star-poor northern winter sky—blue-white Spica high in the north in Virgo the Maiden, and yellow-white Arcturus, lower in the sky in Boötes the Herdsman. A third star, Regulus, shines over in the northwest in Leo the Lion. Between these three stars lies the North Galactic Pole, a region poor in stars but rich in galaxies beyond our Milky Way. Hundreds of galaxies, too faint to see with the unaided eye, await telescopic exploration in the constellations of Leo, Coma Berenices, and Virgo.

Southern Hemisphere
June ■ Looking South

The southern Milky Way is at its best, stretching from horizon to horizon across the south. In its midst stands Crux the Southern Cross, due south and upright in the sky. To the right, or west, of Crux look for the hazy patch of the Eta Carinae Nebula flanked by several bright clusters of stars. To the east of Crux, Alpha and Beta Centauri, also known as Rigel Kentaurus and Hadar, shine high in the south. Between them and Crux, look for a "hole" in the Milky Way, the Coalsack, a dark cloud of interstellar dust 500 light-years away.

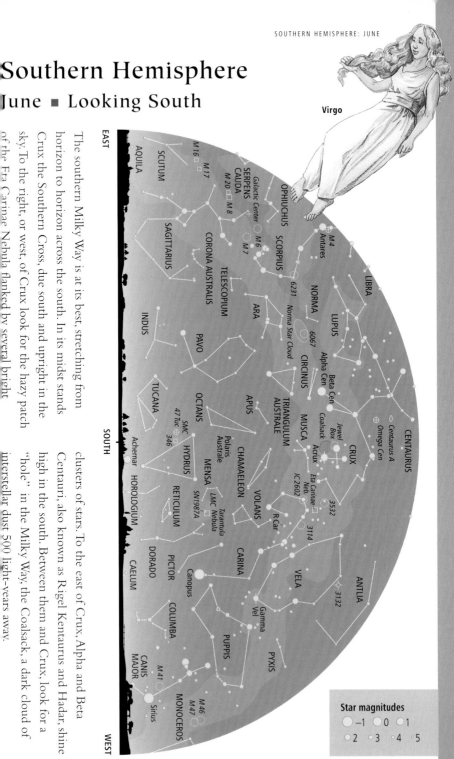

Virgo

EAST

SOUTH

WEST

AQUILA
SCUTUM
M 16
M 17
SERPENS CAUDA
M 20
M 8
OPHIUCHUS
Galactic Center
M 6
M 7
SCORPIUS
M 4
Antares
LIBRA
SAGITTARIUS
CORONA AUSTRALIS
6231
NORMA
Norma Star Cloud
6067
LUPUS
TELESCOPIUM
ARA
CIRCINUS
Alpha Cen
Beta Cen
CENTAURUS
INDUS
PAVO
Coalsack
Jewel Box
Centaurus A
Omega Cen
TUCANA
OCTANS
APUS
TRIANGULUM AUSTRALE
MUSCA
Acrux
CRUX
SMC
47 Tuc
346
Polaris Australe
Eta Carinae Neb.
3532
Achernar
HYDRUS
MENSA
CHAMAELEON
IC 2602
3114
HOROLOGIUM
RETICULUM
LMC
Tarantula Nebula
SN1987A
VOLANS
R Car
DORADO
PICTOR
CARINA
VELA
3132
ANTLIA
CAELUM
Canopus
Gamma Vel
COLUMBA
PUPPIS
PYXIS
CANIS MAJOR
M 41
M 46
M 47
MONOCEROS
Sirius

Star magnitudes
○ −1 ○ 0 ○ 1
○ 2 · 3 · 4 · 5

Southern Hemisphere
July ■ Looking North

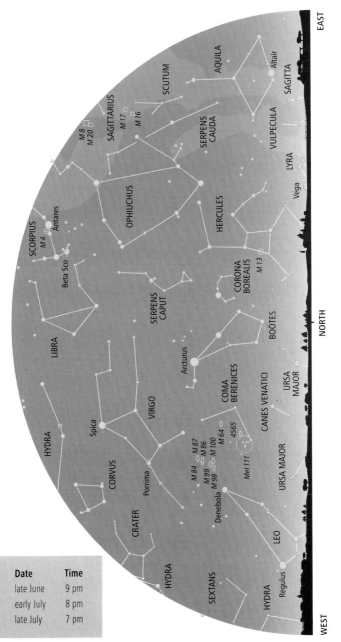

EAST

NORTH

WEST

Date	Time
late June	9 pm
early July	8 pm
late July	7 pm

The brightest star in the north is Arcturus in Boötes the Herdsman, the third brightest in the sky; after Sirius and Canopus. To the right, or east, of Arcturus a neat semi-circle of stars forms Corona Borealis, the Northern Crown. Above Arcturus shines fainter Spica in Virgo the Maiden, while high overhead ruddy Antares gleams as the heart of Scorpius the Scorpion. From Scorpius the Milky Way flows down into the northeastern sky through the constellations of Sagittarius the Archer, Scutum the Shield, and Aquila the Eagle.

Southern Hemisphere
July ▪ Looking South

The Milky Way and the constellations of Crux the Southern Cross and Centaurus the Centaur begin their descent into the west. Rising high into the east is the "fish-hook" of stars that forms Scorpius the Scorpion, followed by a "teapot" of stars, Sagittarius the Archer.

From a dark site, look for a giant "emu" formed by dark lanes in the Milky Way. The Coalsack near Crux forms his head, a dark band from Alpha Centauri to Scorpius forms his neck, while his legs are found in dark lanes streaming down from Scorpius.

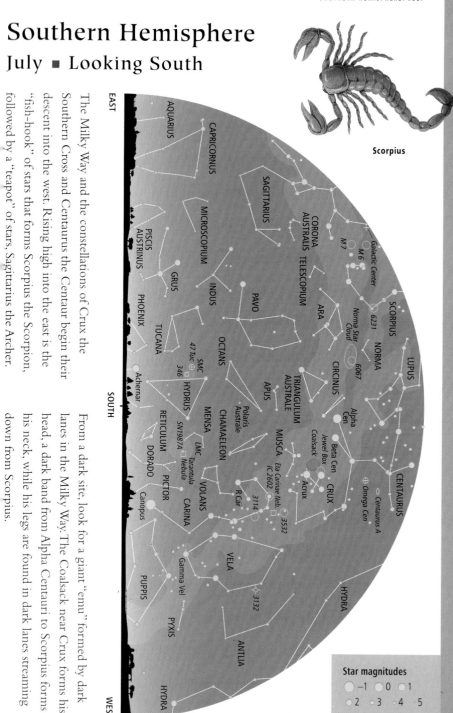

Scorpius

EAST

SOUTH

WEST

AQUARIUS
CAPRICORNUS
SAGITTARIUS
MICROSCOPIUM
PISCIS AUSTRINUS
GRUS
PHOENIX
CORONA AUSTRALIS
TELESCOPIUM
INDUS
PAVO
TUCANA
ARA
NORMA
SCORPIUS
LUPUS
CIRCINUS
TRIANGULUM AUSTRALE
APUS
OCTANS
HYDRUS
RETICULUM
MENSA
CHAMAELEON
MUSCA
CENTAURUS
CRUX
Acrux
VOLANS
CARINA
VELA
PUPPIS
PYXIS
ANTLIA
HYDRA
DORADO
PICTOR
Canopus
Achernar
SMC
47 Tuc
346
LMC
SN 1987A
Tarantula Nebula
R Car
Eta Carinae Neb.
IC 2602
3114
3532
Gamma Vel
3132
M7
M6
Galactic Center
6231
Norma Star Cloud
6067
Alpha Cen
Beta Cen
Jewel Box
Coalsack
Omega Cen
Centaurus A
Polaris Australe

Star magnitudes
−1 0 1
2 3 4 5

301

Southern Hemisphere
August ▪ Looking North

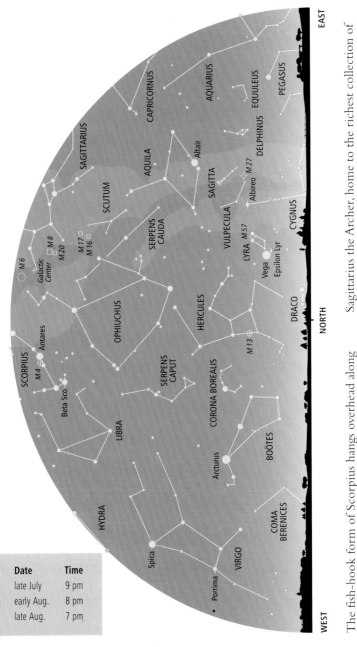

EAST

NORTH

WEST

Date	Time
late July	9 pm
early Aug.	8 pm
late Aug.	7 pm

The fish-hook form of Scorpius hangs overhead along with Antares, the reddish heart of the Scorpion. Straight up lies the center of the Milky Way Galaxy and dozens of star clusters and nebulas perfect for binocular scanning. Sweep down into the north and you will pass through Sagittarius the Archer, home to the richest collection of clusters and nebulas in the sky. The Milky Way narrows and dims below Sagittarius as it passes through Aquila the Eagle, marked by the bright star Altair. Just off the Milky Way shines blue–white Vega.

Southern Hemisphere
August ■ Looking South

It may be mid-winter but this is a superb time to travel to a dark site. The panorama of our Galaxy now sweeps across the heavens from horizon to horizon. Directly overhead lies the center of the Galaxy. There, the Milky Way is at its brightest and widest, filled with rich star clouds and dark lanes that provide marvelous fields for scanning with binoculars—just lie back and enjoy the view. To the south of Scorpius the Scorpion, the Milky Way narrows but remains spectacular through the starfields of Lupus the Wolf, Norma the Square, and Ara the Altar.

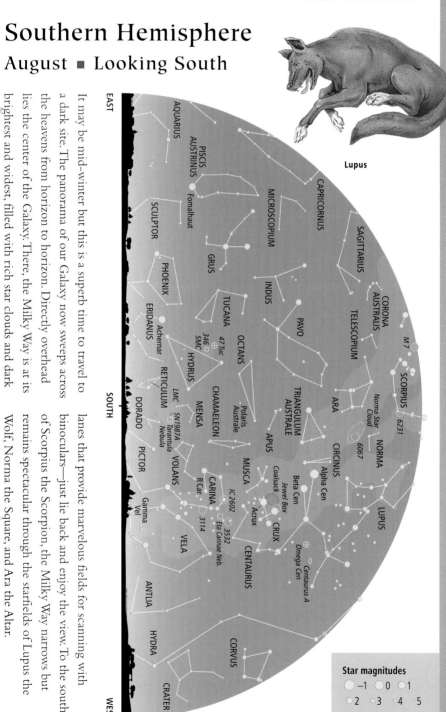

Lupus

EAST

SOUTH

WEST

AQUARIUS
PISCIS AUSTRINUS
Fomalhaut
SCULPTOR
CAPRICORNUS
MICROSCOPIUM
SAGITTARIUS
CORONA AUSTRALIS
M7
SCORPIUS
6231
Norma Star Cloud
NORMA
6067
LUPUS
TELESCOPIUM
ARA
CIRCINUS
Alpha Cen
Beta Cen
Centaurus A
Omega Cen
PHOENIX
GRUS
INDUS
PAVO
TRIANGULUM AUSTRALE
APUS
MUSCA
Jewel Box
Coalsack
Acrux
CRUX
CENTAURUS
ERIDANUS
Achernar
TUCANA
346
SMC
47 Tuc
OCTANS
Polaris Australe
MENSA
CHAMAELEON
CARINA
R Car
IC 2602
3532
Eta Carinae Neb.
3114
VELA
Gamma Vel
ANTLIA
HYDRA
CRATER
CORVUS
HYDRUS
RETICULUM
LMC
DORADO
SN1987A
Tarantula Nebula
PICTOR
VOLANS

Star magnitudes
−1 0 1
2 3 4 5

Southern Hemisphere
September ■ Looking North

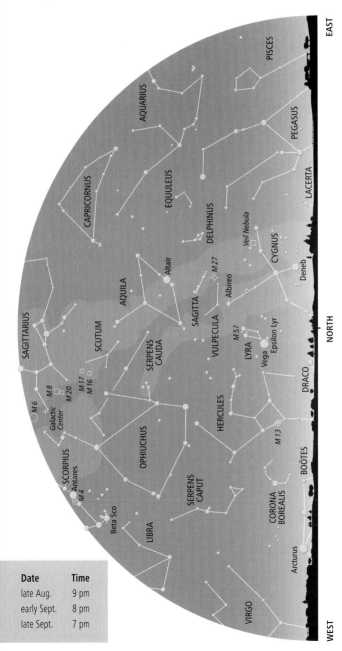

EAST

PISCES

PEGASUS

AQUARIUS

LACERTA

EQUULEUS

CAPRICORNUS

DELPHINUS

Veil Nebula

M 27

Altair

CYGNUS

Albireo

Deneb

AQUILA

SAGITTA

SAGITTARIUS

SCUTUM

VULPECULA

M 57

LYRA

Vega

NORTH

SERPENS
CAUDA

Epsilon Lyr

M 17
M 16

M 8

M 20

M 6

Galactic
Center

HERCULES

DRACO

OPHIUCHUS

M 13

SCORPIUS

Antares

M 4

SERPENS
CAPUT

BOÖTES

Beta Sco

CORONA
BOREALIS

LIBRA

Arcturus

VIRGO

WEST

Date	Time
late Aug.	9 pm
early Sept.	8 pm
late Sept.	7 pm

The Milky Way bisects the northern sky as it flows down from Sagittarius the Archer, through Aquila the Eagle and Cygnus the Swan. Three bright stars dominate the northern sky—Altair in Aquila, Vega in Lyra the Harp, and just above the horizon, Deneb in Cygnus. Northern

Hemisphere observers know these stars as the Summer Triangle. Cygnus is also better known as the Northern Cross, now due north on the horizon. It is much larger than the Southern Cross and, from southern latitudes, appears to be upside down.

Southern Hemisphere
September ■ Looking South

The Milky Way is magnificent, flowing from Sagittarius the Archer overhead, through the star-filled constellations of Scorpius the Scorpion, Lupus the Wolf, Norma the Square, Ara the Altar, Centaurus the Centaur, and Crux the Southern Cross. But as spring begins, these migrate toward the west, leaving the eastern half of the sky sparse by comparison. East of the celestial pole shines Achernar, the mouth of Eridanus the River and the brightest star in that half of the sky. Above, and alone in a star-poor region, shines Fomalhaut in Piscis Austrinus, the Southern Fish.

Eridanus

EAST

SOUTH

WEST

CETUS

SCULPTOR

AQUARIUS

Fomalhaut

FORNAX

PISCIS AUSTRINUS

PHOENIX

GRUS

ERIDANUS

MICROSCOPIUM

SAGITTARIUS

HOROLOGIUM

Achernar

INDUS

RETICULUM

TUCANA

346

47 Tuc

SMC

HYDRUS

OCTANS

PAVO

TELESCOPIUM

CORONA AUSTRALIS

DORADO

LMC

SN1987A

Tarantula Neb.

Polaris Australe

ARA

M7

VOLANS

MENSA

CHAMAELEON

PICTOR

CARINA

R Car.

APUS

TRIANGULUM AUSTRALE

Norma Star Cloud

6231

NORMA

SCORPIUS

IC 2602

3114

MUSCA

3532

Coalsack

Beta Cen

CIRCINUS

6067

Eta Carinae Neb.

Acrux

Jewel Box

Alpha Cen

VELA

CRUX

Eta Carinae

CENTAURUS

Omega Cen

Centaurus A

LUPUS

HYDRA

CORVUS

Spica

VIRGO

○ −1 ○ 0 ○ 1
○ 2 · 3 · 4 5

Southern Hemisphere
October ■ Looking North

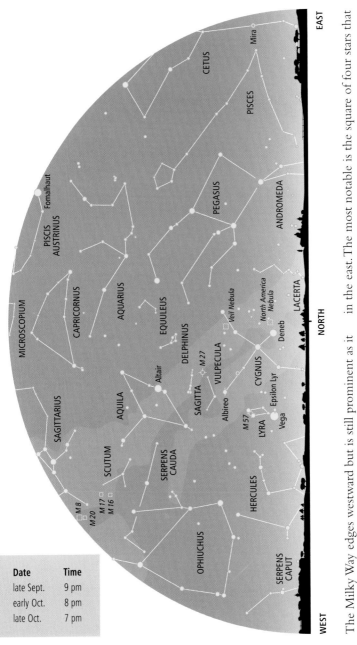

Date	Time
late Sept.	9 pm
early Oct.	8 pm
late Oct.	7 pm

The Milky Way edges westward but is still prominent as it cascades from the star clouds of Sagittarius the Archer to flow between the triangle formed by Altair, Vega, and Deneb, just skimming the northern horizon. As the Milky Way retreats to the west, new spring constellations appear in the east. The most notable is the square of four stars that marks Pegasus the Flying Horse. Above lie dim Aquarius the Water Bearer, Capricornus the Sea Goat, Cetus the Sea Monster, Pisces the Fish, and Piscis Austrinus, the Southern Fish, marked by bright Fomalhaut.

Southern Hemisphere
October ■ Looking South

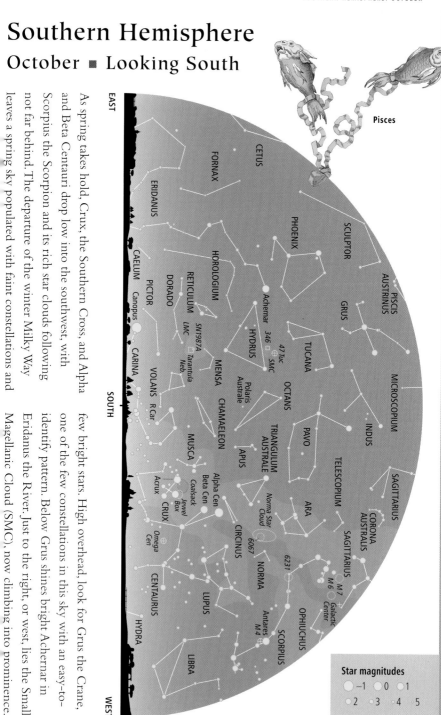

Pisces

EAST

SOUTH

WEST

As spring takes hold, Crux, the Southern Cross, and Alpha and Beta Centauri drop low into the southwest, with Scorpius the Scorpion and its rich star clouds following not far behind. The departure of the winter Milky Way leaves a spring sky populated with faint constellations and

few bright stars. High overhead, look for Grus the Crane, one of the few constellations in this sky with an easy-to-identify pattern. Below Grus shines bright Achernar in Eridanus the River. Just to the right, or west, lies the Small Magellanic Cloud (SMC), now climbing into prominence.

CETUS
FORNAX
ERIDANUS
CAELIUM
PICTOR
Canopus
CARINA
DORADO
RETICULUM
HOROLOGIUM
PHOENIX
SCULPTOR
PISCIS
AUSTRINUS
GRUS
MICROSCOPIUM
SAGITTARIUS
Achernar
346
LMC
SN1987A
Tarantula
Neb.
HYDRUS
47 Tuc
SMC
TUCANA
OCTANS
PAVO
INDUS
TELESCOPIUM
CORONA
AUSTRALIS
SAGITTARIUS
VOLANS R Car
MENSA
Polaris
Australe
CHAMAELEON
TRIANGULUM
AUSTRALE
APUS
ARA
M7
M6
Galactic
Center
MUSCA
Alpha Cen
Beta Cen
Norma Star
Cloud
6067
6231
OPHIUCHUS
Acrux
Coalsack
Jewel
Box
CRUX
Omega
Cen
CIRCINUS
NORMA
Antares
M4
SCORPIUS
CENTAURUS
LUPUS
LIBRA
HYDRA

Star magnitudes
−1   0   1
2   3   4   5

Southern Hemisphere
November ■ Looking North

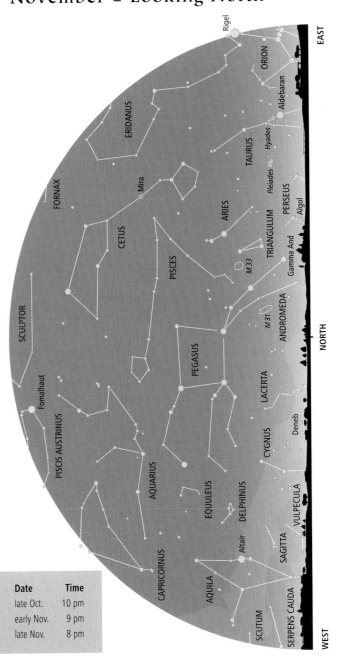

WEST

NORTH

EAST

Rigel

ORION

Aldebaran

TAURUS

Hyades

Pleiades

PERSEUS

Algol

ARIES

TRIANGULUM

Gamma And

ANDROMEDA

M33

M31

PISCES

LACERTA

CETUS

Mira

FORNAX

ERIDANUS

PEGASUS

CYGNUS

Deneb

SCULPTOR

Fomalhaut

PISCIS AUSTRINUS

AQUARIUS

EQUULEUS

DELPHINUS

VULPECULA

SAGITTA

AQUILA

Altair

CAPRICORNUS

SCUTUM

SERPENS CAUDA

Date	Time
late Oct.	10 pm
early Nov.	9 pm
late Nov.	8 pm

The Milky Way lies all around you on the horizon. In spring you gaze out the bottom of the Galaxy, at right angles to its disk of stars that forms the Milky Way. We are looking through the thinnest part of the Galaxy, so this sky contains few bright stars. Across the north stretches a series of faint constellations: Cetus the Sea Monster, Pisces the Fish, Aquarius the Water Bearer, Capricornus the Sea Goat, and Piscis Austrinus, the Southern Fish. Around the south galactic pole in Sculptor, and in nearby Fornax the Furnace, lie dozens of galaxies ripe for telescopic viewing.

Southern Hemisphere
November ■ Looking South

A celestial aviary populates the southern sky. Apus, the Bird of Paradise, flits low in the southwest. Above struts Pavo the Peacock, while higher still flies Grus the Crane. High in the south perch Tucana the Toucan and Phoenix, the mythological bird reborn from the ashes of its predecessor. All these constellations were invented by 16th-century Dutch navigators. Other faint constellations named for scientific instruments (Telescopium, Micro-scopium, Octans the Octant, Horologium the Clock, Pictor the Compass) date from the 18th century.

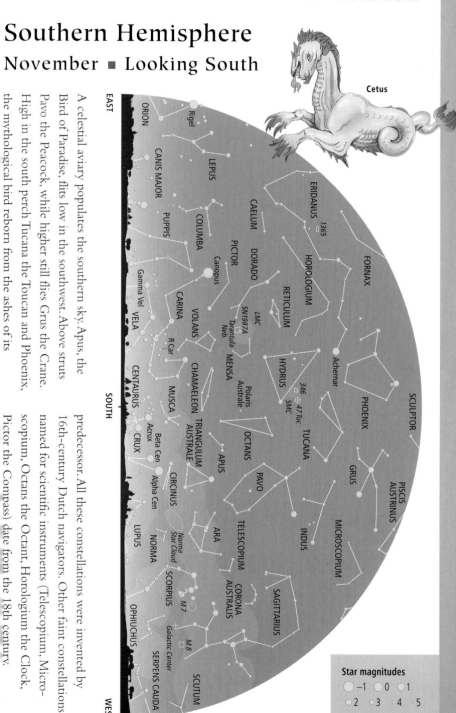

Cetus

EAST

SOUTH

WEST

ORION
Rigel
CANIS MAJOR
LEPUS
PUPPIS
COLUMBA
CAELUM
ERIDANUS
1365
HOROLOGIUM
FORNAX
SCULPTOR
PISCIS AUSTRINUS
Gamma Vel
VELA
CARINA
PICTOR
Canopus
DORADO
RETICULUM
Achernar
PHOENIX
GRUS
MICROSCOPIUM
R Car
VOLANS
LMC
SN1987A
Tarantula Neb.
MENSA
HYDRUS
346
47 Tuc
SMC
TUCANA
INDUS
CENTAURUS
MUSCA
CHAMAELEON
Polaris Australe
OCTANS
APUS
PAVO
TELESCOPIUM
SAGITTARIUS
Acrux
CRUX
Beta Cen
TRIANGULUM AUSTRALE
CIRCINUS
ARA
CORONA AUSTRALIS
Alpha Cen
LUPUS
NORMA
Norma Star Cloud
SCORPIUS
M7
Galactic Center
M8
SCUTUM
OPHIUCHUS
SERPENS CAUDA

Star magnitudes
−1   0   1
2   3   4   5

309

Southern Hemisphere
December ■ Looking North

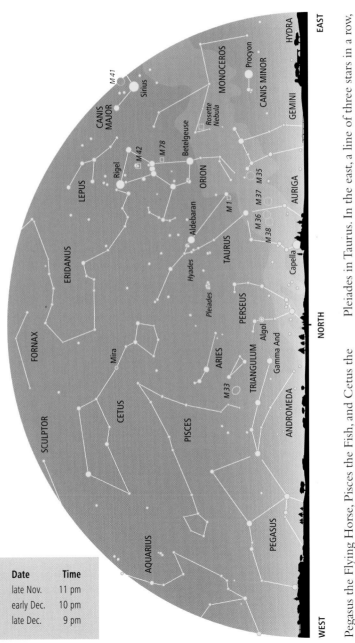

EAST

NORTH

WEST

Date	Time
late Nov.	11 pm
early Dec.	10 pm
late Dec.	9 pm

Pegasus the Flying Horse, Pisces the Fish, and Cetus the Sea Monster head off into the west, to be replaced by the brighter patterns of Taurus the Bull, Orion the Hunter, and Canis Major, the Great Dog, now rising in the east. Nearly due south lies an eye-catching cluster of stars, the Pleiades in Taurus. In the east, a line of three stars in a row, unique in the sky, marks the Belt of Orion. Compared to how he is traditionally depicted, Orion stands upside down in the Southern Hemisphere sky, with his head and shoulders below the Belt and his legs above.

Southern Hemisphere
December ▪ Looking South

Taurus

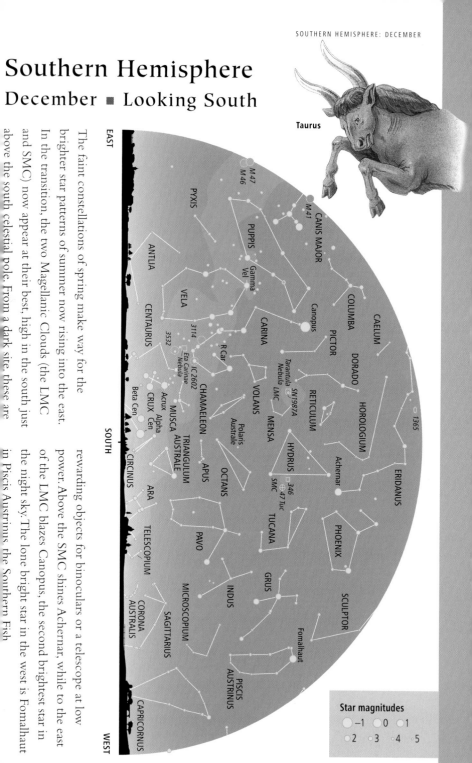

The faint constellations of spring make way for the brighter star patterns of summer now rising into the east. In the transition, the two Magellanic Clouds (the LMC and SMC) now appear at their best, high in the south just above the south celestial pole. From a dark site, these are rewarding objects for binoculars or a telescope at low power. Above the SMC shines Achernar, while to the east of the LMC blazes Canopus, the second brightest star in the night sky. The lone bright star in the west is Fomalhaut in Piscis Austrinus, the Southern Fish

EAST

SOUTH

WEST

M 47
M 46
M 41
CANIS MAJOR
PYXIS
PUPPIS
ANTLIA
Gamma Vel
VELA
3114
3532
CENTAURUS
R Car
IC 2602
Eta Carinae Nebula
Beta Cen
Acrux
CRUX Alpha Cen
MUSCA
CHAMAELEON
TRIANGULUM AUSTRALE
CIRCINUS
ARA
TELESCOPIUM
CORONA AUSTRALIS
SAGITTARIUS
MICROSCOPIUM
CAPRICORNUS
PISCIS AUSTRINUS
Fomalhaut
INDUS
GRUS
PAVO
OCTANS
APUS
Polaris Australe
VOLANS
MENSA
HYDRUS
SMC
346
47 Tuc
TUCANA
SCULPTOR
PHOENIX
Achernar
RETICULUM
Tarantula Nebula LMC
SN1987A
CARINA
PICTOR
Canopus
COLUMBA
DORADO
HOROLOGIUM
CAELUM
ERIDANUS
1365
1

Star magnitudes
○ −1 ○ 0 ○ 1
○ 2 · 3 · 4 · 5

Sky
Tours

You are about to embark on a series of guided tours of the cosmos. The simple directions and easy-to-use star charts in this chapter will show you a treasure trove of celestial objects. With a little practice, you will soon be hopping from starfield to cluster to galaxy to nebula.

"Whereas other animals hang their heads and look at the ground, he made man stand erect, bidding him look up to heaven, and lift his head to the stars."

Ovid
Roman poet, 43 BC–AD 17?

Starhopping the Sky

The technique of starhopping uses key star patterns as celestial signposts to guide you to the best and brightest wonders of the deep sky.

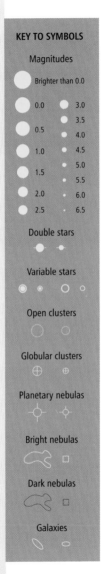

KEY TO SYMBOLS

Magnitudes

Brighter than 0.0

0.0   3.0

             3.5

0.5   4.0

             4.5

1.0   5.0

1.5   5.5

2.0   6.0

2.5   6.5

Double stars

Variable stars

Open clusters

Globular clusters

Planetary nebulas

Bright nebulas

Dark nebulas

Galaxies

Starhopping is the number-one method experienced stargazers use to find telescopic targets. For the beginner, the 10 starhops on the following pages are a great way to start out, by taking you on guided tours of some fascinating regions. Use the charts and the accompanying directions to "hop" to dozens of the best double and variable stars, star clusters, nebulas, and galaxies.

Basically, starhopping uses bright stars as a guide to finding fainter objects. First, note a bright star or star pattern near the target. Center that pattern in your binoculars or telescope's finderscope. From that guidepost hop over to the position of the target. You will find that binoculars and finderscopes have fields of view about twice as wide as the "spotlight" circles around each object.

Icons indicate the "tools" needed to see the object.

👁 naked eye

⌕ binoculars

⌕ telescope

Objects marked with multiple icons can be seen with all the methods listed but will be best in the highest power tool.

Each map shows stars as faint as magnitude 6.5, the limit for unaided eyes. All deep-sky objects visible in a small telescope are plotted.

Grid lines of right ascension and declination are plotted in purple. Constellation boundaries are in black. The "join-the-dot" constellation patterns are shown as light gray lines.

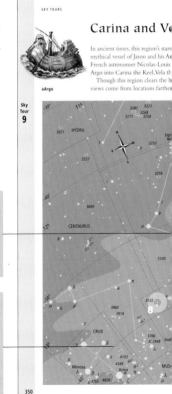

Carina and Ve

In ancient times, this region's stars
mythical vessel of Jason and his Ar
French astronomer Nicolas-Louis
Argo into Carina the Keel, Vela th
Though this region clears the h
views come from locations farther

aArgo

Sky
Tour
9

e 6 Comae. Now switch to the main telescope. Just west
within the same low-power eyepiece field, look for a sliver
s M 98, a spiral galaxy lying edge-on. Move an eyepiece
to find M 99, a spiral galaxy seen face-on. M 100,
on spiral, lies about two eyepiece fields (2 degrees)
Comae. This trio serves as your entrance to the Coma-
luster. For M 100, see also p. 400.

M 86 (NGC 4374 and 4406) Recenter 6
ead southeast again to M 99. Now keep going the same
er two eyepiece fields beyond M 99. Across the border in
I see a pair of elliptical galaxies, M 84 and M 86, glowing
de. From here, you can travel northeast along the sky's
ne of galaxies, Markarian's Chain—a dozen systems like
ee arc.

C 4486 Recenter M 84 and M 86. Now
degrees southeast. Look for a single bright glow. This is
rmous elliptical galaxy at the heart of the Coma-Virgo
o p. 402.

renices Cluster (Mel 111) Look for a faint
about 14 degrees northeast of Denebola. Visible to

unaided eyes on a dark night, this cluster, known officially as Melotte
111, makes a fine sight in binoculars. It appears large because it is
nearby, only 270 light-years away.

6 Porrima (Gamma [γ] Virginis) Use high power to resolve
this tight double star into celestial headlights, a pair of equally matched
yellow-white stars. As the stars orbit each other, they are coming closer
together. For a few months in 2008 they will be impossible to separate.

M 99
Lying about 50 million
light-years away in the
Coma-Virgo cluster is
the spiral galaxy M 99.
Through a telescope, it
appears almost circular
in shape.

Sky Tour 2

M 100
At 7 arcminutes across,
M 100 presents the
largest apparent size
of any galaxy in the
Coma-Virgo cluster. It
is relatively easy to see
because of its brilliant
starlike core.

1 Canopus Unlike nearby and apparently brighter Sirius,
Canopus, the sky's second brightest star, is distant and truly
luminous—313 light-years away and 13,000 times brighter
than the Sun.

2 Gamma [γ] Velorum Binoculars split this wide
double star, with magnitude 1.8 and 4.3 components set in a
rich Milky Way field.

3 NGC 3132, the Eight Burst Nebula The size of the
northern sky's Ring Nebula, this bright planetary contains a

Seen best in
Northern Hemisphere winter*
Southern Hemisphere summer
*only from south of 25°N latitude

Sky Tour 9

325

Photos depict some of
the best objects on each
tour. Remember, though,
that through your
telescope's eyepiece you
will see less detail than
these long exposures,
usually taken with much
larger scopes, can reveal.

This box indicates the
best time of year (for
both Northern and
Southern Hemispheres)
to see this region of sky
in the evening.

351

The 10 sky tours take in the most interesting celestial regions, containing the sky's best targets. Each map covers an area 60 degrees wide, or roughly six fist-widths held at arm's length (see Measuring the Sky, p. 92). Within that area, each tour contains 6 to 10 highlighted targets, the best of dozens plotted.

The tours take you to nearly every type of celestial object—colorful double stars, pulsating variable stars, sparkling star clusters, glowing bright nebulas, dusty dark nebulas, and ghostly galaxies. Tour 1 even contains a star—47 Ursae Majoris— known to have a planet orbiting it.

The sky tours are ordered according to the times that they come into view in the evening

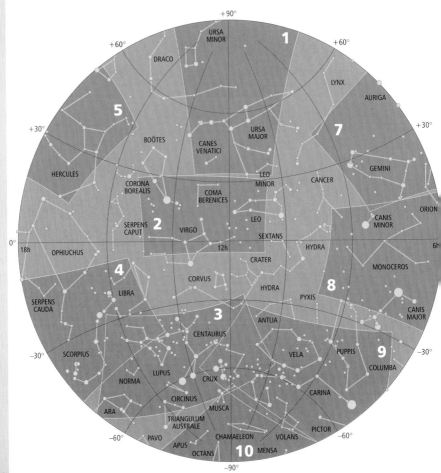

throughout the year. Tours 1, 2, and 3 are best undertaken in March to May. Tours 4 and 5 are for June to September. Tour 6 is for October to December, while tours 7, 8, and 9 are all best in the period January to April. Tour 10 is suitable for any season, with the exception of midwinter in the Southern Hemisphere.

Although most tours can be observed anywhere, tour 1 is mainly for northern observers, and three tours (3, 9, and 10) are primarily, if not exclusively, for observers south of the equator. Tour 2 (Virgo and Coma) takes in the sky's richest gathering of distant galaxies. While this tour highlights a half dozen of the best galaxies, many more await the ardent observer.

Now that you know the basics of starhopping, you are ready to undertake any of the sky tours featured on pp. 318–54.

Seeing Red
The star charts' colors have been chosen to show up well when viewed with a red-light flashlight, highly recommended for preserving your night vision.

Use these key charts to locate the regions of sky covered by the 10 detailed Sky Tours maps. The maps are represented by the dark blue patches.

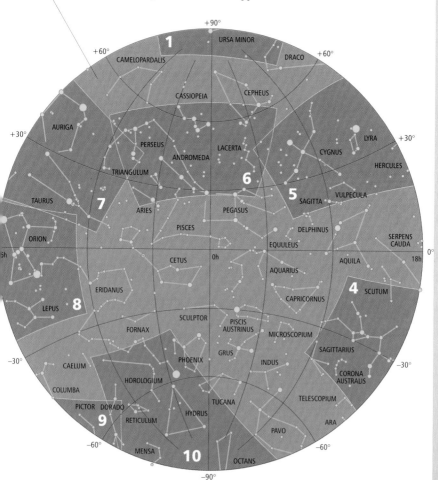

The North Polar Sky

This northern sky tour focuses mainly on the constellations Ursa Major and Ursa Minor, and offers skywatchers plenty of celestial treasures. In Greek mythology, Ursa Major and Ursa Minor are the beautiful Callisto and her son Arcas, changed into bears by a jealous Hera, then flung into the heavens and immortalized by Hera's husband—and Callisto's lover—Zeus, chief of the gods.

Ursa Major

Sky
Tour
1

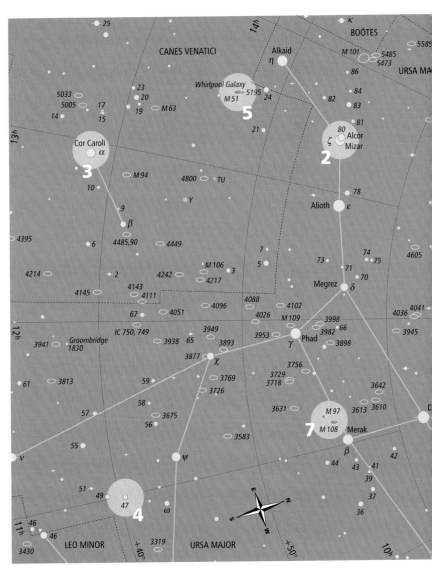

1 Polaris, The North Star (Alpha [α] Ursae Minoris)

👁 📷 Polaris is a Cepheid variable star 46 times larger than our Sun and 432 light-years away. Through a telescope, look for a faint (magnitude 9.0) pale blue companion star 18 arcseconds away from Polaris.

2 Mizar and Alcor (Zeta [ζ] and 80 Ursae Majoris)

👁 📷 📷 Look at Mizar, the middle star in the Big Dipper's handle, with your unaided eyes. You should be able to make out a faint star, Alcor, just above it. A pair of binoculars splits them

Seen best in

Northern Hemisphere spring

Southern Hemisphere *

* not visible south of 10°S latitude

Sky Tour 1

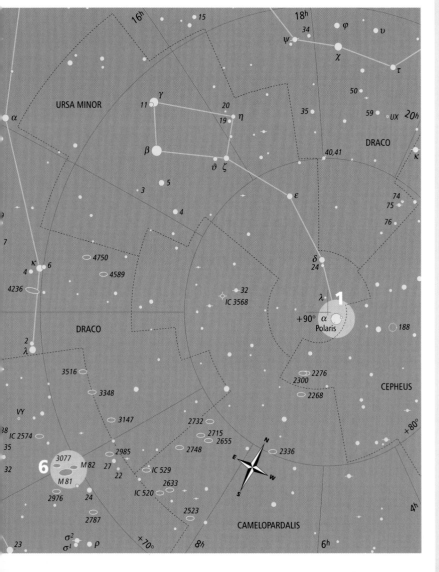

Alcor and Mizar
Below right: Alcor (left)
and Mizar (right) make
up the apparent wide
double star in the Big
Dipper's handle. Mizar
is itself a double star—
its companion can just
be made out here.

Polaris
Below: The relatively
faint stars of Ursa
Minor appear to swing
around Polaris (top)
as the sky rotates
during the night.

easily, but a telescope reveals Mizar has another companion, a magnitude 4 star, Mizar B, just 14 arcseconds away.

3 Cor Caroli (Alpha [α] Canum Venaticorum) Named to honor King Charles II, Cor Caroli (The Heart of Charles) splits easily in any telescope into a magnitude 2.9 blue-white gem with a fainter magnitue 5.5 yellow-white jewel 20 arcseconds away.

4 47 Ursae Majoris In 1996, astronomers discovered this yellow Sun-like star has a planet with a mass of at least 2.6 Jupiters (so it is likely a gas giant) orbiting it every three years.

5 M 51, the Whirlpool Galaxy (NGC 5194) First described by the Earl of Rosse in 1845, the Whirlpool is arguably the sky's finest

spiral galaxy. Visible in binoculars as a hazy spot, an 8 inch (200 mm) telescope begins to show the spiral arms well. See also p. 406.

6 M 81 (NGC 3031) and M 82 (NGC 3034) 🔭🔭 Draw an imaginary diagonal line from Gamma [γ] to Alpha [α] Ursae Majoris, and then extend that line an equal distance north. In binoculars, look for two small gray patches. In a telescope, the spiral galaxy M 81 shows as an elliptical glow, while the irregular galaxy M 82 looks like a mottled spindle of light. For M 81, see also p. 398.

7 M 108 (NGC 3556) and M 97, the Owl Nebula (NGC 3587) 🔭 A much fainter deep sky duo lies below the Big Dipper bowl near Merak (Beta [β] Ursae Majoris). M 97, the Owl Nebula, is a nearby planetary nebula, the cast-off shell of an aging star. See also p. 378. A low-power field away lies M 108, a distant spiral galaxy seen edge-on.

M 81 and 82
In telescopes, a low-power eyepiece nicely frames a duo of bright galaxies: M 82 (top) and M 81 (bottom).

Virgo and Coma Berenices

Coma Berenices

Thousands of galaxies are scattered throughout Virgo and Coma Berenices. This mighty club of distant systems is known as the Coma-Virgo galaxy cluster. Here, you are looking at right angles to the disk of stars that forms the Milky Way. Our Galaxy's north pole lies in Coma Berenices, named for the flowing tresses of Queen Berenice who ruled Egypt with her husband Ptolemy Euergetes in the 3rd century BC.

Sky Tour 2

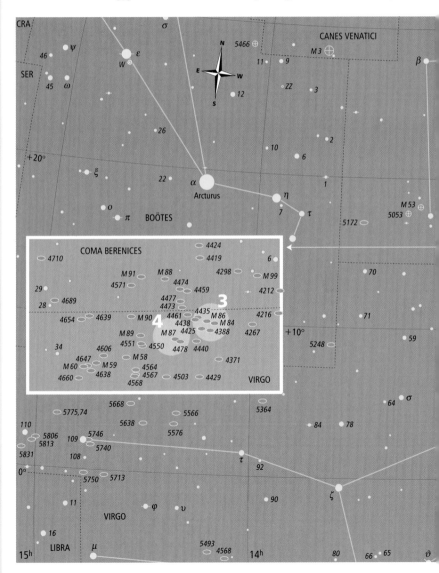

1 Denebola (Beta [β] Leonis) 👁 ᴄᴏᴏ Denebola's name comes from the Arabic *dhanab al-asad*, meaning the Lion's Tail, a reference to its position as the tuft of hair on the tail of Leo. This magnitude 2.1 blue-white star lying 36 light-years away serves as the starting point for a tour of galaxies millions of light-years distant.

2 M 98, M 99, M 100 (NGC 4192, 4254, 4321) ᴄᴏ From Denebola, swing your telescope one finderscope field (6.5 degrees) due east and center the next brightest star you find,

Seen best in
Northern Hemisphere spring
Southern Hemisphere autumn

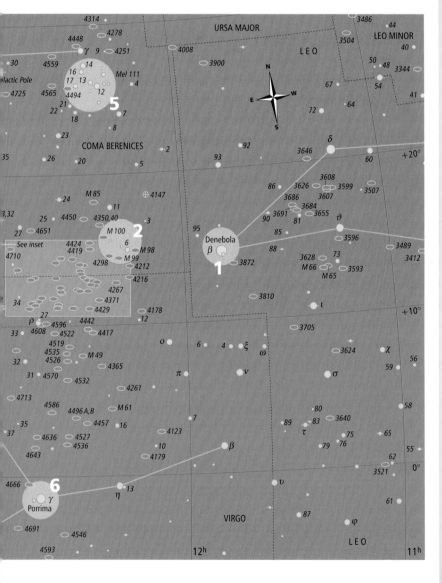

5th-magnitude 6 Comae. Now switch to the main telescope. Just west of 6 Comae, within the same low-power eyepiece field, look for a sliver of light. This is M 98, a spiral galaxy lying edge-on. Move an eyepiece field southeast to find M 99, a spiral galaxy seen face-on. M 100, another face-on spiral, lies about two eyepiece fields (2 degrees) northeast of 6 Comae. This trio serves as your entrance to the Coma-Virgo galaxy cluster. For M 100, see also p. 400.

3 M 84 and M 86 (NGC 4374 and 4406) Recenter 6 Comae, then head southeast again to M 99. Now keep going the same way, for another two eyepiece fields beyond M 99. Across the border in Virgo, you will see a pair of elliptical galaxies, M 84 and M 86, glowing at 9th magnitude. From here, you can travel northeast along the sky's richest gathering of galaxies, Markarian's Chain—a dozen systems lie along a 2-degree arc.

4 M 87 (NGC 4486) Recenter M 84 and M 86. Now hop about 1.5 degrees southeast. Look for a single bright glow. This is M 87, the enormous elliptical galaxy at the heart of the Coma-Virgo cluster. See also p. 402.

5 Coma Berenices Cluster (Mel 111) Look for a faint cluster of stars about 14 degrees northeast of Denebola. Visible to

Markarian's Chain
Virgo harbors an amazing sight for the telescope user: a great "river" of galaxies, Markarian's Chain. In this view, the chain stretches from the lower right up to the bright "headlights" of M 84 and M 86, at center left. M 87 is the large elliptical galaxy at top right.

unaided eyes on a dark night, this cluster, known officially as Melotte 111, makes a fine sight in binoculars. It appears large because it is nearby, only 270 light-years away.

6 Porrima (Gamma [γ] Virginis) ⊂○ Use high power to resolve this tight double star into celestial headlights, a pair of equally matched yellow-white stars. As the stars orbit each other, they are coming closer together. For a few months in 2008 they will be impossible to separate.

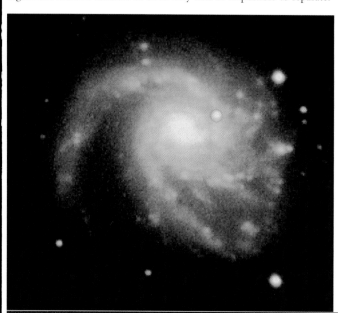

M 99

Lying about 50 million light-years away in the Coma-Virgo cluster is the spiral galaxy M 99. Through a telescope, it appears almost circular in shape.

M 100

At 7 arcminutes across, M 100 presents the largest apparent size of any galaxy in the Coma-Virgo cluster. It is relatively easy to see because of its brilliant starlike core.

Sky Tour 2

325

Centaurus and Crux

Through the constellations Centaurus and Crux the Milky Way wanders to its most southerly point. Here, southern stargazers are treated to fine views of a number of interesting and visually stunning celestial objects. Northern Hemisphere observers might like to consider a trip south, for these skies harbor some of the very best examples of many classes of deep–sky objects.

Centaurus

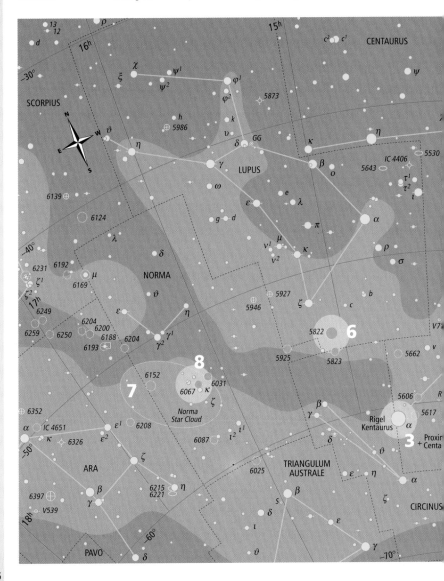

1 NGC 4755, the Jewel Box Cluster 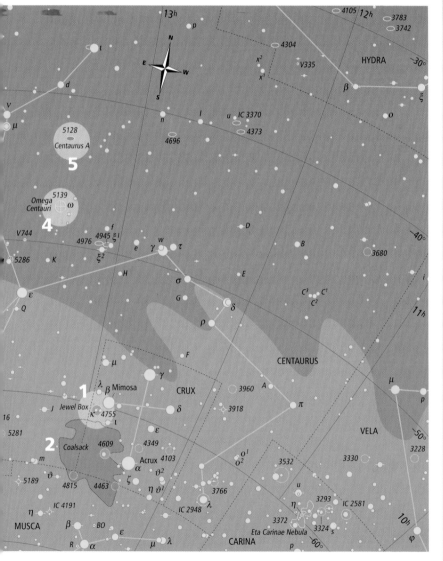 Near Mimosa, or Beta [β] Crucis, in Crux, this collection of blue stars contrasting with a lone red star reminded the 19th-century astronomer John Herschel of sparkling jewelry. See also p. 370.
2 The Coalsack Just below the Jewel Box lies what seems to be a large hole in the Milky Way, best seen with the unaided eye or binoculars. A cloud of interstellar dust 600 light-years away blocks distant starlight, creating this Coalsack, the sky's best dark nebula.

Seen best in
Northern Hemisphere spring *
Southern Hemisphere autumn

* from south of 20°N latitude

Sky
Tour
3

327

3 Rigel Kentaurus (Alpha [α] Centauri)  The nearest bright star to our Solar System is also one of the sky's best double stars. A telescope splits Alpha Centauri into two stars 16 arcseconds apart. The brighter of the two is a twin of the Sun in color and temperature.

4 Omega [ω] Centauri 👁 ⊂OO ⊂O Visible to the unaided eye and cataloged as a star, this fuzzy spot resolves in any telescope into the sky's finest globular cluster. Thousands of stars provide the impression of sparkling spilled sugar. A stunning sight. See also p. 374.

5 Centaurus A (NGC 5128) ⊂OO ⊂O A finder field (4 degrees) north of Omega Centauri lies one of the brightest but oddest galaxies. Centaurus A's elliptical glow is bisected by a dark dust lane, perhaps the

Centaurus A
Right: A dark band of dust cuts across Centaurus A, a strong source of radio energy.

Coalsack
Below: The dark region in the center of the photo, standing out like a hole among the star clouds of the Milky Way, is the Coalsack, one of the largest and densest dark nebulas in the sky. Its closest neighbors are Crux and the Jewel Box.

remains of a galaxy collision. Both Omega Centauri and Centaurus A can be seen from southern U.S. latitudes.

6 NGC 5822 Across the border in Lupus lies this large, relatively rich (with a hundred stars) open cluster, best observed in binoculars or through a telescope at low power.

7 Norma Star Cloud A pair of binoculars reveals this brilliant patch of Milky Way in the constellation Norma as a superb field of uncountable stars.

8 NGC 6067 Within the star-packed Norma Star Cloud lies an even denser cluster of stars, NGC 6067. Binoculars show it, but a low-power telescope reveals one of the sky's best open clusters set amid a rich Milky Way starfield.

Norma Star Cloud
Midway along the Milky Way's glittering path in this photo is the Norma Star Cloud, indicated by the circle. The two bright stars toward the bottom right corner are Rigel Kentaurus and Hadar.

Sky
Tour
3

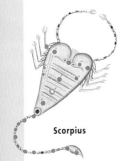

Scorpius

Scorpius and Sagittarius

Scorpius and Sagittarius cover a particularly rich and varied part of the night sky. When you look up at these constellations you will be gazing toward the very center of our Milky Way Galaxy. The actual core lies hidden behind dark clouds of interstellar dust. But between us and the core lie the numerous starfields, nebulas, and star clusters that adorn the entire region.

Sky Tour 4

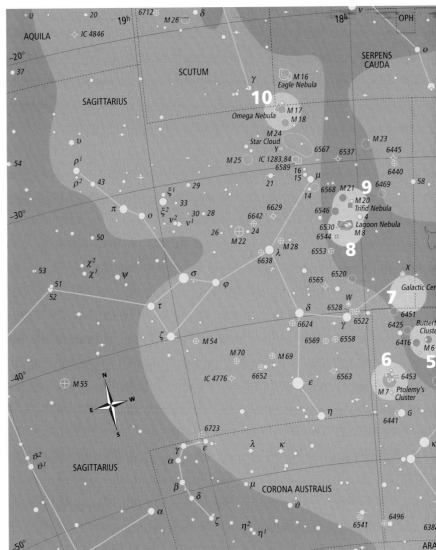

1 Antares (Alpha [α] Scorpii) 👁 📷 Bearing a Greek name that means "the rival of Mars," this red supergiant star appears yellow-orange to the unaided eye. Antares is roughly 300 times the size of the Sun, so big that if it were placed at the center of our Solar System, it would engulf all of the planets out to the orbit of Mars.

2 M 4 (NGC 6121) 📷 📷 Little more than a low-power eyepiece field west of Antares glows M 4, a "squashed" globular cluster, an effect caused by the object's rapid rotation.

Seen best in
Northern Hemisphere summer
Southern Hemisphere winter

Sky
Tour
4

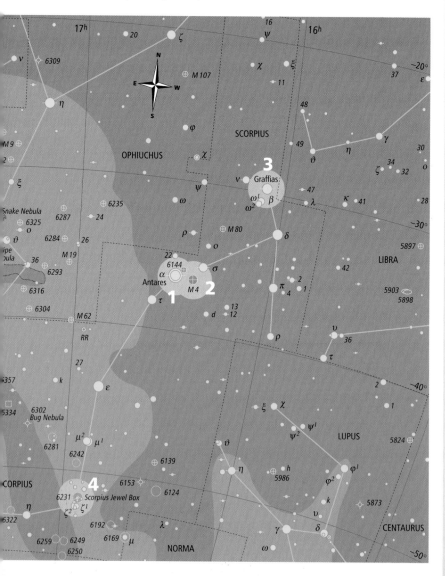

3 Graffias (Beta [β] Scorpii)  In a telescope, this naked-eye star splits into a magnitude 2.6 star with a dimmer magnitude 4.9 companion, both tinted blue-white.

4 NGC 6231, the Scorpius Jewel Box Visible to the unaided eye as a hazy spot in the Scorpion's tail, binoculars reveal a large, loose collection of 120 stars.

5 M 6, the Butterfly Cluster (NGC 6405) Thought to resemble a butterfly with outstretched wings, this sprawling open cluster lies set against a dark, dusty region of the Milky Way.

6 M 7, Ptolemy's Cluster (NGC 6475) Recorded by Claudius Ptolemy more than 1,800 years ago, this naked-eye open cluster outdoes M 6 for richness and size. Use ultra-low power to take in its 1-degree diameter.

7 Galactic Center North of M 6 and M 7, the Milky Way brightens into a dense field of faint stars, the Sagittarius Star Cloud, toward the Galactic Center. In binoculars, rivers and fingers of inky dust lanes give the impression of canyons cutting through the stars.

Antares

The brightest star in Scorpius is Antares, its brilliance and yellow-orange color making it unmistakable. This red supergiant is thought to be about 600 light-years away.

8 M 8, the Lagoon Nebula (NGC 6523) ⊂◯◯ ⊂◯ A binocular field north of the main Star Cloud glows an elongated, grayish patch of light crossed by a dark lane, with a loose star cluster, NGC 6530, on one side of this dark "lagoon." See also p. 382.

9 M 20, the Trifid Nebula (NGC 6514) ⊂◯ Two telescope fields north of the Lagoon, look for a compact glow trisected by three dark lanes that give this nebula its name. See also p. 380.

10 M 17, the Omega or Swan Nebula (NGC 6618) ⊂◯◯ ⊂◯ In a dark sky, this glowing gas cloud resembles a swimming swan, the Greek letter Omega, or a number 2. See also p. 386.

M 6
Left: Also known as the Butterfly Cluster, M 6 is a 5th-magnitude open cluster lying 1,500 light-years from Earth.

M 7
Below: Binoculars reveal many of the 80 or so stars in the bright open cluster M 7, also known as Ptolemy's Cluster. A wide-field eyepiece should be used for a good view through a telescope.

Cygnus, Lyra, and Hercules

Around and between the stars of the Summer Triangle lies a favorite hunting ground for Northern Hemisphere stargazers on warm summer nights (though this region is also visible from south of the equator, appearing low in the austral winter sky). Star clusters, large "wide-field" nebulas, and two of the sky's best planetary nebulas highlight this part of the northern Milky Way.

Cygnus

Star chart showing Cygnus, Lyra, Hercules region with labeled stars and deep-sky objects:

LAC — M39, IC 1369, 7026, 51, 20h, ψ, 20, 7, κ
W, ρ, 7062, 63, 59, ω², 43, 26, 6826, 17, ϑ, ι
IC 5117, 7039, 60, ω¹, U, o², 30, 6811
75, North America Nebula, 7000, 55, α, o¹
77, 76, 68, ξ, 57, 56, Deneb, 1, δ
74, 7027, IC 5067,70, Pelican Nebula, 6866, 14
79, 72, ν, 6914, V973, 6819
σ, 6910, γ
70, 69, τ, 61, CYGNUS, 40, M29, P, 22, 19, η, ϑ
υ, λ, 44, 42, IC 4996, 29, 28, 25, 15, 11
τ, 47, 36, 27, 6871, 17, 8, 4
ε, 35, η, χ
ζ, 6992, Veil Nebula, 49, 39
B, 6995, 6960, 52, 48, φ, 9, 2, M56
32, T, 6940, 41, 6834, β, Albireo
31, 21, 23, 15, VULPECULA, 4
27, 26, 19, 18, 20, 6885,82, 10, 3
30, 28, 25, 24, 16, 13, 8, α
33, 22, 17, 6830, 14, 6823, 2
PEG, 29, Dumbbell Nebula, M27, 5, 1
U, EU, 6905, 18, ϑ, 6886, η, The Coathanger, Cr 399
7006, DELPHINUS, γ, M71, ζ, α, 9, SAGITTA
γ, α, IC 4997, 6879, 13, δ, β
δ, 11

1 Deneb (Alpha [α] Cygni) 👁 Start your Triangle tour with Deneb, a blue-white supergiant star that shines 250,000 times brighter than our Sun. At 3,200 light-years away, Deneb is one of the most distant stars visible to the naked eye.

2 NGC 7000, the North America Nebula Within a binocular field southeast of Deneb glows the form of the North America Nebula. Visible to the unaided eye as a bright patch, this object becomes an obvious sight in a telescope at very low power equipped with a nebula filter.

Seen best in
Northern Hemisphere summer
Southern Hemisphere winter

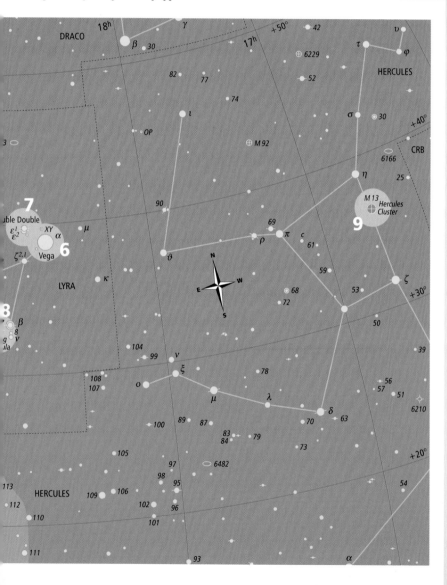

3 NGC 6960 and 6992/5, the Veil Nebula One strand of this supernova remnant runs through the star 52 Cygni. The other arc lies 2.5 degrees east. Though detectable in binoculars, the two arcs look best through a large telescope with a nebula filter.

4 Albireo (Beta [β] Cygni) This showpiece double features a magnitude 3.1 golden yellow star with a magnitude 5.1 sapphire blue companion. See also p. 360.

See also p. 360.

Cygnus
Right: The Northern Cross in Cygnus lies astride the bright northern Milky Way. Deneb is at the top of the photo; Vega, in Lyra, shines at far right.

Sky Tour 5

NGC 7000
Below: Photos show NGC 7000 to look like the shape of North America—hence its common name—but this resemblance is not readily apparent to the eye when observing.

5 M 27, the Dumbbell Nebula (NGC 6853)  Between Albireo and Altair, and 3 degrees due north of the "arrowhead" star of Sagitta, glows the Dumbbell, one of the largest and brightest planetary nebulas in the sky. See also p. 390.

6 Vega (Alpha [α] Lyrae) 👁 The most westerly Summer Triangle star is a blue-white giant 25 light-years from us.

7 Epsilon [ε] Lyrae (the Double Double) Use Vega to hop the short distance to Epsilon Lyrae. Your finderscope will show it as a double, but at high telescopic power each star resolves into two stars about 2 arcseconds apart. See also p. 360.

8 M 57, the Ring Nebula (NGC 6720) Between Beta [β] and Gamma [γ] Lyrae lies a tiny gray smoke ring, an easy target for any telescope. This cast-off shell from a faint, dying star is the sky's best-known planetary nebula. See also p. 388.

9 M 13, the Hercules Cluster (NGC 6205) No tour of the Summer Triangle is complete without a visit to the best globular cluster in the northern sky. A 4 inch (10 cm) telescope begins to resolve this swarm of stars into countless pinpoints. See also p. 372.

Sky
Tour
5

Vega
Vega is a sparkling blue-white gem among a scattering of much duller jewels. It is the fifth brightest star in the entire sky.

Andromeda and Cassiopeia

As the legend has it, Queen Cassiopeia angered Poseidon, god of the seas, with her vanity. To appease the divine wrath, husband King Cepheus ordered their daughter, Andromeda, be sacrificed to the jaws of Cetus the Sea Monster. Perseus rescued Andromeda, turning Cetus to stone by holding up the head of the horrid Medusa.

Left: Perseus rescuing Andromeda

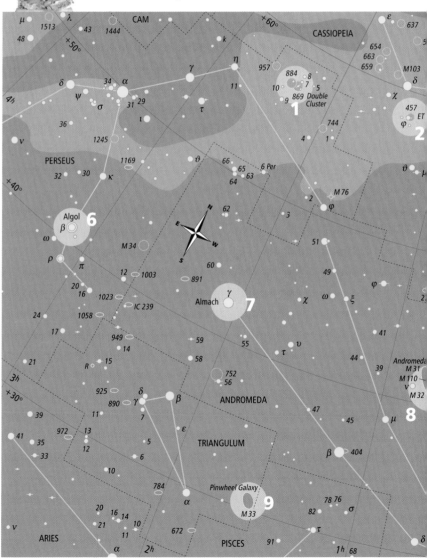

1 NGC 869 and 884, the Double Cluster 👁 ᑕ◯◯ ᑕ◯
A naked-eye fuzzy patch in the head of Perseus resolves into
twin star clusters in binoculars and two wide splashes of stars
filling a telescope eyepiece. See also p. 368.
2 NGC 457, the ET Cluster ᑕ◯ Easy to find, this sparse
cluster is a delight—some see an owl in its stars, others liken its
shape to ET, the Extraterrestrial.
3 NGC 7789 ᑕ◯ A sprinkling of 300 stars in an area roughly
the apparent size of the Moon makes NGC 7789 one of the

Seen best in
Northern Hemisphere autumn
Southern Hemisphere spring*

*only from north of 15°S latitude

Sky
Tour
6

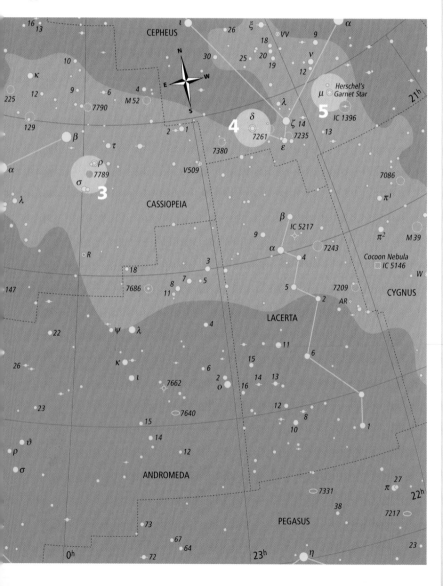

sky's richest open clusters. It forms a right-angle triangle with the stars Alpha [α] and Beta [β] Cassiopeiae.

4 Delta [δ] Cephei 👁 ⊂∞ ⊂∞ A telescope splits this star into a fine pale orange and blue double. The yellow star is the prototype for the Cepheid variables, yellow supergiants that regularly pulsate in size and vary in brightness.

5 Mu [μ] Cephei, Herschel's Garnet Star ⊂∞ ⊂∞ Appearing in small telescopes as one of the reddest stars in the sky, this cool supergiant is so large it would engulf Jupiter if it replaced our Sun.

6 Algol, the Demon Star (Beta [β] Persei) 👁 ⊂∞ This naked-eye star dims from magnitude 2.1 to 3.4 for 10 hours every 2 days, 21 hours. A dimmer star orbiting and eclipsing Algol causes the winking of this evil eye of the demon Medusa. See also p. 362.

7 Almach (Gamma [γ] Andromedae) ⊂∞ Described as gold and blue, the stars of this double present one of the best color contrasts of the sky's many binary star systems.

NGC 457
More than a hundred
stars pack the open
cluster NGC 457.

8 M 31, the Andromeda Galaxy (NGC 224) 👁 ⫷ΟΟ ⫷Ο
Some 2.5 million light-years away, M 31 is the most distant object
easily visible to the unaided eye. Binoculars show it best, as an
elongated glow. A telescope reveals two companion galaxies, M 32
(NGC 221) and M 110 (NGC 205). See also p. 408.

9 M 33, the Pinwheel Galaxy (NGC 598) ⫷ΟΟ ⫷Ο Below
Andromeda lies another large and nearby spiral galaxy, but one
fainter and more diffuse than M 31. Haze or light pollution may
render M 33 invisible. See also p. 410.

NGC 7789
Left: An impressive
half a degree across,
NGC 7789 is a
magnitude 6.7 open
cluster in Cassiopeia.

M 33
Below: The Pinwheel
Galaxy is notoriously
tricky to spot—you will
need a very dark night
and a viewing site well
away from city lights.
The extra effort may
be rewarded by a view
of its loosely wound
spiral structure.

Taurus, Auriga, and Gemini

Taurus

Just as Sagittarius and Scorpius lie toward the center of our Milky Way Galaxy, Taurus, Auriga, and Gemini lie in exactly the opposite direction. Look to where these three constellations converge and you look toward the near edge of our Galaxy. Whether using binoculars or a telescope, observers will find rich pickings here among the bright stars and brilliant clusters.

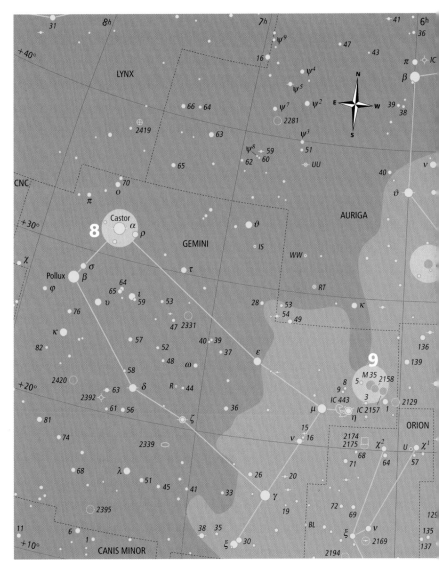

1 Aldebaran (Alpha [α] Tauri) 👁 Yellow-orange to the unaided eye, this cool giant star seems set amid the Hyades star cluster. Its membership is just an illusion. Aldebaran lies 65 light-years away, only half the distance to the Hyades.

2 The Hyades (Melotte 25) 👁 ᴄᴏᴏ This V-shaped cluster of 200 stars is so close (150 light-years) that it appears 10 times wider than the Moon. Binoculars frame the cluster best.

3 M 45, the Pleiades (Melotte 22) 👁 ᴄᴏᴏ ᴄᴏ The finest and most famous open star cluster, the Pleiades is an obvious

Seen best in
Northern Hemisphere winter
Southern Hemisphere summer

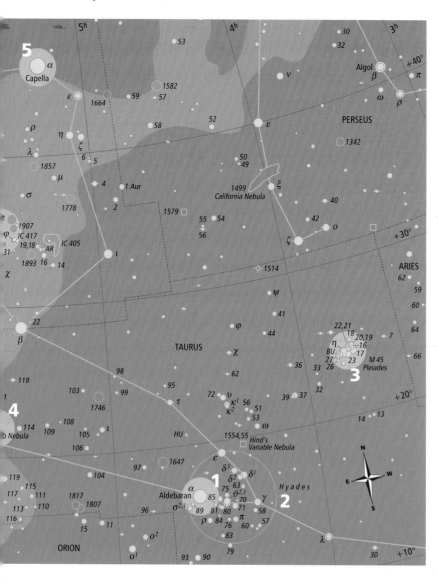

naked-eye sight and a wonderful target for binoculars or a low-power telescope. Dozens of young, blue-white stars populate this newly formed cluster 400 light-years away. See also p. 366.

4 M 1, the Crab Nebula (NGC 1952)  This fuzzy gray patch in a telescope represents the other end of the stellar lifecycle from the Pleiades' young stars. The Crab is the expanding debris of a supernova star that was seen to explode in AD 1054. See also p. 392.

The Hyades

Right: The stars of the Hyades cluster form a distinctive V shape— the bull's head in Taurus—around yellow-orange Aldebaran, which marks the bull's eye.

Capella

Below: At magnitude 0.08, Capella is the brightest star in Auriga and the sixth brightest in the sky. Its four components are too close together to be separated through regular telescopes.

5 Capella (Alpha [α] Aurigae) 👁 Shining from across 42 light-years, Capella is actually a pair of yellow giants, each with a companion star, that orbit too close to resolve in any telescope eyepiece.

6 M 36 (NGC 1960) and M 38 (NGC 1912) 🔭 🔭 Only 2 degrees apart, these two star clusters form a contrasting pair: with M 36 smaller and brighter, and M 38 larger but fainter and more scattered.

7 M 37 (NGC 2099) 🔭 🔭 Just beyond the east side of the pentagon-shaped Auriga glows this rich cluster of several hundred stars, the finest of Auriga's three Messier clusters.

8 Castor (Alpha [α] Geminorum) 👁 🔭 The brightest star of Gemini is a close double, with components only 4 arcseconds apart that take 470 years to orbit one another. Each is also a binary, too close to be seen as separate stars.

9 M 35 (NGC 2168) 🔭 🔭 At a distance of 2,800 light-years, this rich star cluster is one of the best in the northern sky. In the same telescopic field look for the small, faint glow of NGC 2158, a star cluster six times farther away than M 35.

Sky Tour
7

M 37, M 36, M 38
Running from lower left to upper right are M 37, M 36, and M 38 against a Milky Way backdrop in Auriga.

Canis Major

Orion and Canis Major

Orion is star-forming territory. Long-exposure photographs and infrared images show that the entire constellation is wrapped in nebulosity—the stuff of new stars. The visible tip of the gas clouds is the Orion Nebula, a star factory some 1,500 light-years away. Trek east of Orion, along the main band of the Milky Way, and you enter star-cluster country.

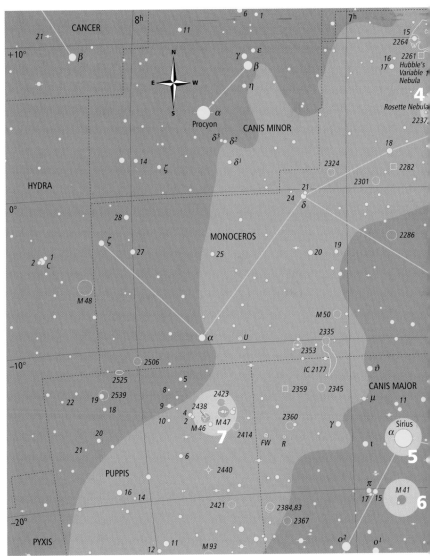

1 Betelgeuse (Alpha [α] Orionis) 👁 This red supergiant's name is a corruption of original Arabic words for "hand of al-jauza," an obscure term which may refer to a female character in ancient Arabic mythology.

2 M 42, the Orion Nebula (NGC 1976) 👁 ⚬⟲ ⟲⚬
Clearly visible to the unaided eye under a dark sky as a fuzzy star in Orion's Sword, M 42 blooms into a wondrous sight in any telescope. Great wings of gas curve away from a glowing mottled core, lit by a quadruplet star, the Trapezium star cluster.

Seen best in
Northern Hemisphere winter
Southern Hemisphere summer

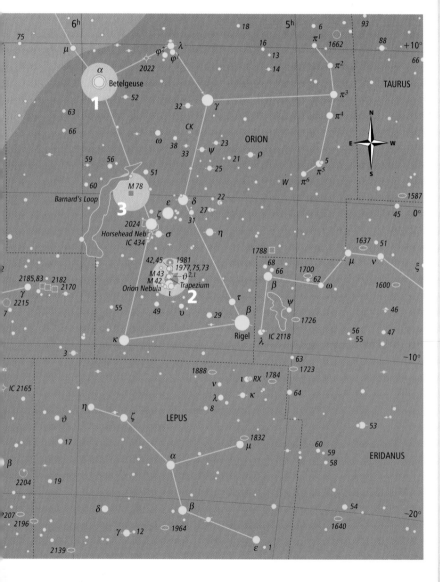

A smaller detached nebula, M 43 (NGC 1982), glows just northeast of the main mass of M 42. See also p. 394.

3 M 78 (NGC 2068)  North of Orion's Belt lies one of the best reflection nebulas. Unlike M 42 which shines by light emitted by fluorescing gas, M 78 shines by reflecting the light of nearby stars.

4 NGC 2244 and the Rosette Nebula (NGC 2237) To the east of Orion, in dim Monoceros, shines a sparse star cluster, NGC 2244, easily visible in binoculars. From a dark site, a telescope equipped with a nebula filter shows a faint wreath of nebulosity surrounding the cluster. Use the lowest power possible.

5 Sirius (Alpha [α] Canis Majoris) Sirius shines so brightly because it is close, only 8.6 light-years away, and because it is a hot, white star over 20 times brighter than our Sun. Its faint companion, Sirius B, the first white dwarf star discovered, now lies too close to blazing Sirius to resolve in most amateur telescopes.

6 M 41 (NGC 2287) To the south of Sirius, well within a binocular or finderscope field, look for a large, bright but sparse open star cluster, M 41.

Orion
Betelgeuse (top left), Rigel (bottom right), and the three-in-a-row Belt stars make Orion an unmistakable sight. Dangling from the belt is one of the marvels of the night sky: M 42, the Orion Nebula.

Sky Tour 8

348

7 M 46 (NGC 2437) and M 47 (NGC2422)  A sizable hop (13 degrees) east of Sirius takes you to a pair of contrasting open star clusters: M 47, a bright but coarse cluster, and nearby M 46, a fainter but richer sprinkling of stars. Within M 46, a telescope shows the tiny disk of the planetary nebula NGC 2438.

M 78
Left: M 78 is a small nebula which glows softly at magnitude 8.3 in reflected light from two young, hot stars embedded within it.

Rosette Nebula
Below: Dark dust lanes give the Rosette Nebula a distinctive appearance. At its center lies the cluster NGC 2244, made up of young stars born from the nebula's gas.

Carina and Vela

In ancient times, this region's stars outlined Argo Navis, the Ship Argo, mythical vessel of Jason and his Argonauts. In the 18th century, the French astronomer Nicolas-Louis de Lacaille subdivided sprawling Argo into Carina the Keel, Vela the Sail, and Puppis the Stern.

Though this region clears the horizon from 25°N latitude, the best views come from locations farther south.

Argo

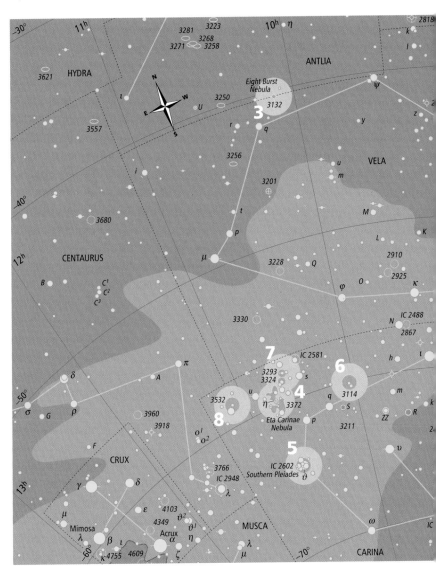

1 Canopus 👁 Unlike nearby and apparently brighter Sirius, Canopus, the sky's second brightest star, is distant and truly luminous—313 light-years away and 13,000 times brighter than the Sun.

2 Gamma [γ] Velorum 👓 👓 Binoculars split this wide double star, with magnitude 1.8 and 4.3 components set in a rich Milky Way field.

3 NGC 3132, the Eight Burst Nebula 👓 The size of the northern sky's Ring Nebula, this bright planetary contains a

Seen best in
Northern Hemisphere winter*
Southern Hemisphere summer

*only from south of 25°N latitude

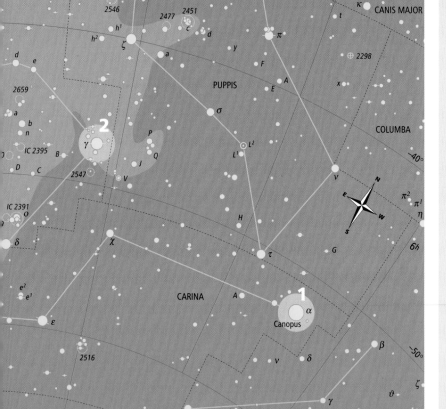

10th magnitude central star, the source of the expanding shell of gas. It lies about 2,600 light-years away.

4 NGC 3372, the Eta Carinae Nebula 👁 ⟨⟩ ⟨⟩ This is the sky's finest nebula, outclassing the Orion Nebula for size. Unaided eyes and binoculars show it, but the best panoramic views of this 4-degree-wide object come in a telescope at low power. Dark dust lanes cross the glowing gas clouds, while embedded in the brightest region of nebulosity shines the remarkable star Eta Carinae. For a few years in the mid-19th century this unstable star flared to outshine Canopus. A high-magnification view of Eta Carinae reveals it surrounded by the golden-yellow Homunculus Nebula, a cloud of debris thrown off in its 19th-century eruption.

NGC 3372

Right: The superb Eta Carinae nebula is one of the sky's greatest sights. Dark rifts made of dust split the nebula into segments.

NGC 3532

Below: Appearing as a a fuzzy patch to the naked eye, NGC 3532 is a magnificent object with binoculars. A telescope of any size will resolve a large number of stars, which are best viewed using low power.

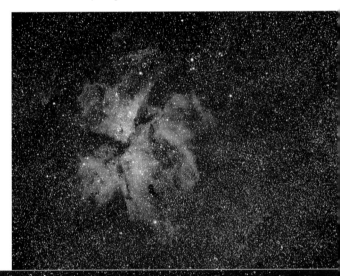

5 IC 2602, the Southern Pleiades  Four prominent star clusters flank the Eta Carinae Nebula. To the south, the brightest is the Southern Pleiades, a bright but coarse cluster surrounding the magnitude 2.7 star Theta [θ] Carinae.

6 NGC 3114 Northeast of the Eta Carinae Nebula, another naked-eye cluster glows at 4th magnitude. NGC 3114 contains about a hundred stars in an area the size of the Moon's disk.

7 NGC 3293 Due north of the Eta Carinae Nebula, NGC 3293 is the smallest of the nebula's flanking clusters. A telescope is best for resolving this tight grouping.

8 NGC 3532 A population of at least 150 faint stars amid an already rich starfield make this elliptically shaped open cluster one of the finest in the sky, north or south.

IC 2602

Southern observers compare IC 2602, with its scattering of bright stars bathed in a blue reflection nebula, to the Pleiades. This open cluster has about 60 members, the brightest being Theta Carinae.

Sky Tour 9

The South Polar Sky

Only stargazers south of the equator see this region well, with its remarkable Clouds of Magellan. These two satellite galaxies of the Milky Way are like little universes, filled with clusters and nebulas numerous enough to occupy nights of exploring.

The clouds were named after the 16th-century Portuguese explorer Ferdinand Magellan, who first described them.

Eridanus

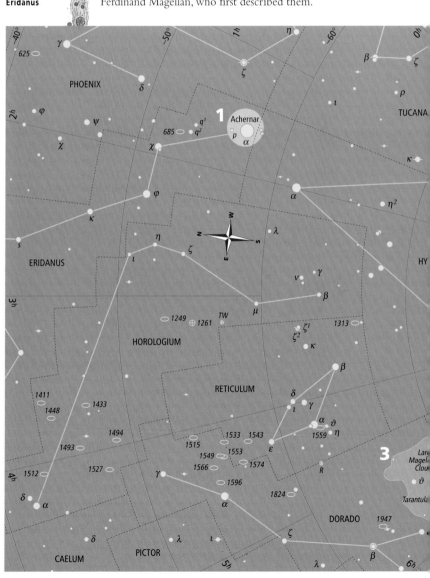

1 Achernar 👁 Achernar shines bright and alone in the south during the austral spring. The star's name stems from Arabic words for "the river's end," from its location at the mouth of Eridanus the River.

2 Sigma [σ] Octantis ⊂○○ This 5th-magnitude star's claim to fame is that it is the closest naked-eye star to the south celestial pole. This is the South Star, or Polaris Australe.

3 Large Magellanic Cloud (LMC) 👁 ⊂○○ ⊂□○ Visible as a diffuse patch to the naked eye, the LMC is the most prominent

Seen best in
Northern Hemisphere
(not visible)
Southern Hemisphere spring

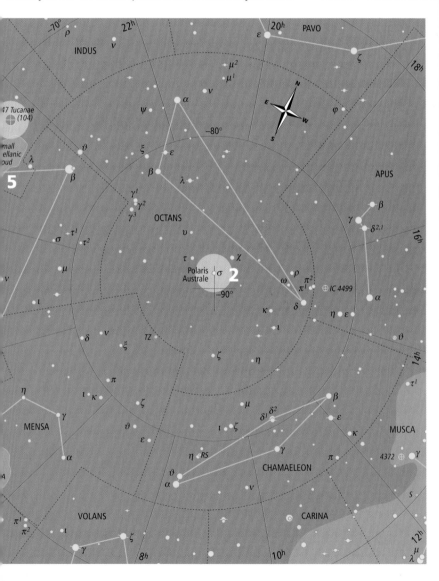

object in the constellation Dorado. Recent estimates place the LMC 163,000 light-years away, making it the closest visible galaxy to the Milky Way. Binoculars neatly frame the LMC, but a scan along its length with a telescope reveals numerous clumps of stars and patches of nebulosity, each with a cataloged NGC number. See also p. 412.

Sky Tour 10

Achernar
Right: Brilliant Achernar stands out in an otherwise barren patch of sky at the "mouth" of Eridanus.

NGC 2070
Below: The wispy tendrils of NGC 2070, also known as the Tarantula Nebula, are clearly seen in this image. An emission nebula, the Tarantula is illuminated by a cluster of hot, young stars at its heart.

4 **NGC 2070, the Tarantula Nebula** ⊂⊙⊙ ⊂⊙ In a telescope, the brightest nebula in the LMC displays a wealth of tentacled structure. The Tarantula Nebula is so large that if it were as close as the Orion Nebula it would span 60 Moon diameters.

5 **Small Magellanic Cloud (SMC)** 👁 ⊂⊙⊙ ⊂⊙ Located mostly in Tucana, the more distant SMC lacks the LMC's array of star-forming nebulas. While the LMC shows hints of a spiral structure, the SMC is clearly an irregular galaxy thanks to its total lack of symmetry. Both galaxies orbit the Milky Way and may be torn apart and absorbed by it in the distant future.

6 **47 Tucanae (NGC 104)** ⊂⊙⊙ ⊂⊙ The SMC's brightest object appears in binoculars as a fuzzy star but resolves in a telescope into a glorious globular cluster. Ranking only second in spectacle after Omega Centauri, 47 Tuc features a densely packed core.

7 **NGC 362** ⊂⊙⊙ ⊂⊙ Overshadowed by neighboring 47 Tucanae, NGC 362 is a fine globular cluster, resolving nicely with a 6 inch (150 mm) telescope. Its location in the SMC is an illusion; NGC 362 lies 28,000 light-years away, placing it within our Galaxy.

SMC
The Small Magellanic Cloud is one of our Galaxy's closest neighbors. In this image, the spectacular globular cluster 47 Tucanae lies to the upper left of the SMC, with the much smaller globular, NGC 362, at lower left.

Sky Tour 10

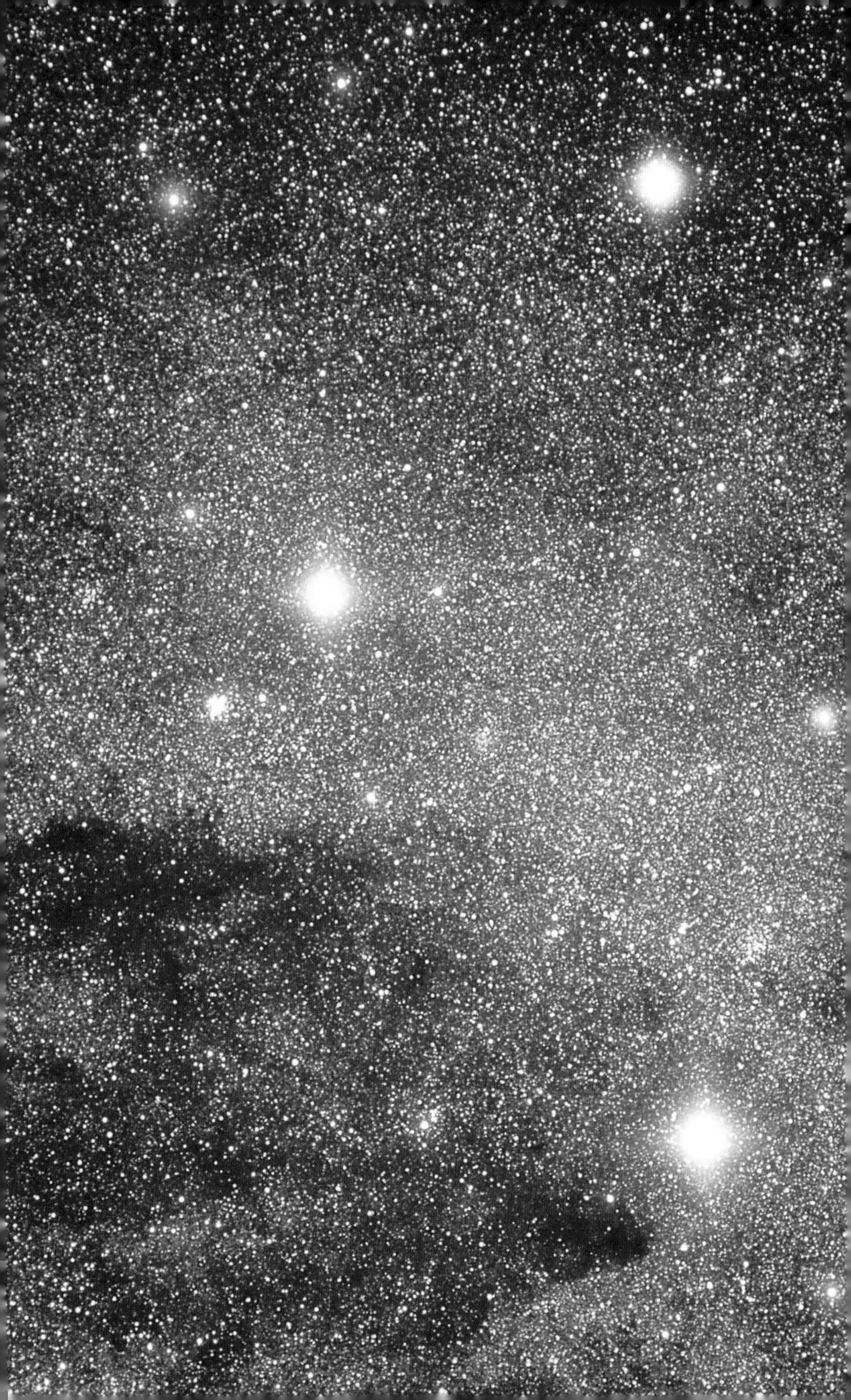

Observing
Stars

Double and multiple stars, variable stars, and glittering star clusters are among the boundless riches of the night sky. This chapter targets a tantalizing selection of these fascinating points of light.

"Stars, I have seen them fall,
But when they drop and die
No star is lost at all
From all the star-sown sky."

A. E. Housman,
English poet, 1859–1936

Double and Multiple Stars

- Constellation: Cygnus (Albireo); Crux (Alpha Crucis); Lyra (Epsilon Lyrae)
- Visibility: Northern Hemisphere and north of 56°S (Albireo);
  Southern Hemisphere and south of 21°N (Alpha Crucis);
  Northern Hemisphere and north of 44°S (Epsilon Lyrae)
- Magnitude: 3.0 (Albireo); 0.75 (Alpha Crucis); 4.5 (Epsilon Lyrae)

Of the millions of stars in the sky, more than half have one or more companion stars. Fortunately for the amateur astronomer, a great many of these are visible in small telescopes.

Double stars can either orbit about a common center of gravity, in which case they are called binary stars, or appear in chance alignments that make them appear associated, in which case they are referred to as optical doubles. Double stars come in an infinite variety of colors, brightnesses, and separations (separation is the apparent distance of the companion, or secondary star, from the brighter star, or primary).

One such study in contrasts is Albireo, which marks the head of Cygnus. Albireo is often said to be the finest double star in the heavens. The primary member is bright yellow, while the fainter component has a bluish tint. The pair's generous separation of 35 arcseconds—twice as wide as Saturn's apparent disk size—makes this an easy object for viewing by low-power telescope.

Alpha (α) Crucis is an example of two stars of nearly the same brightness and color, but with only a hair's separation between them. This star—13th brightest in the night sky— marks the foot of Crux. Its components are tinted blue-white and have a separation of only 4 arcseconds.

Double and triple stars are fairly common. In rare instances, however, double stars come with multiple members. One famous example is Epsilon (ε) Lyrae, known as the "Double Double." Epsilon lies northeast of Vega, the brightest star in the constellation Lyra. A 4 inch (100 mm) telescope at medium magnification splits Epsilon into two stars with a separation of 208 arcseconds (or 3.4 arcminutes). Increased magnification, however, reveals that each component itself has a companion separated by a little over 2 arcseconds, making this a challenging but exceptional quadruple star for the small telescope.

Other bright double stars with visible companions include Rigel (in Orion), Algeiba (Leo), Mizar (Ursa Major), Polaris (Ursa Minor), and Castor (Gemini).

Albireo
Whether you are observing this star from the dark of the country or the middle of a city, Albireo is one of the prettiest sights in the night sky. One of its components is golden yellow in color with a magnitude of 3, and the other is bluish with a magnitude of 5.

Epsilon Lyrae
A small telescope or a pair of binoculars easily shows the double nature of Epsilon Lyrae, as here. A larger telescope transforms the star into a spectacular quadruple system.

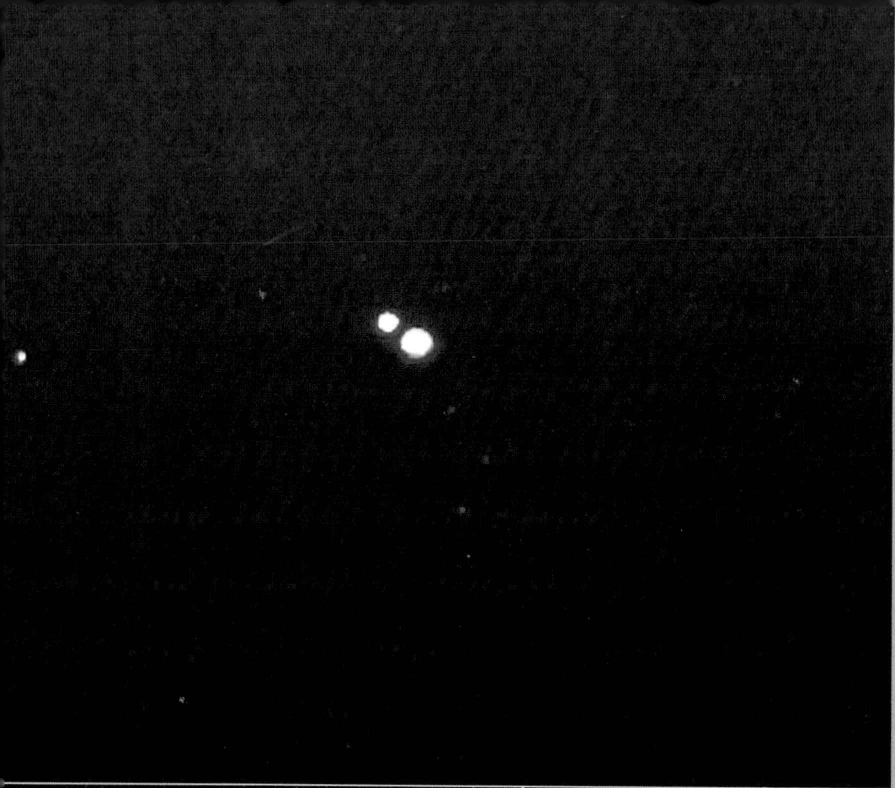

Observing Variable Stars

- Constellation: Perseus (Algol); Cetus (Mira)
- Visibility: Northern Hemisphere and north of 42°S (Algol); both hemispheres (Mira)
- Magnitude variation: 2.1 to 3.4 (Algol); 3 to 10 (Mira)
- Period: 2.8 days (Algol); 331 days (Mira)

Tens of thousands of stars vary in brightness—some dramatically in a matter of days, others more gradually over a period of years. Like double stars, variables offer skywatchers a wide assortment; each has its own unique attributes.

A small number of variable stars may be observed with the naked eye. Delta (δ) Cephei in Cepheus and Beta (β) Lyrae in Lyra are two such examples, each varying by about one magnitude over a period of 5 and 13 days, respectively. Binoculars and telescopes bring vastly more variables within reach of skywatchers.

Beta (β) Persei, better known as Algol, is a good place for the budding variable star observer to begin. Algol is classified an eclipsing variable. Every 2 days, 21 hours, a fainter but slightly larger companion star passes between it and Earth, causing Algol to drop from magnitude 2.1 to 3.4. By comparing its brightness to those of adjacent stars from night to night—many of which can be seen in binoculars or a low-power telescope—these changes become readily apparent.

Another fascinating variable is Omicron (o) Ceti, better know as Mira, in Cetus. Mira is a dying star whose outer gaseous shell expands and contracts as its internal heat flow fluctuates. At its faintest, Mira falls between magnitudes 8 and 10; at its brightest between magnitudes 3 and 4. On rare occasions, however, it has become as bright as magnitude 1. Patience is a virtue with observing this star, however, because it takes about 331 days to cycle through its brightness changes. About a dozen Mira-type variables are easily visible at their peak brightness.

Pulsating Mira
Mira is the brightest long-period pulsating variable. It may go from magnitude 10 at minimum (above) to magnitude 3 at maximum (below). This pattern was first noticed in 1596 by the Dutchman David Fabricius.

Winking Algol
The star that winks, Algol, is the most famous of the eclipsing binaries. The regular drop in brightness, caused by a larger star passing in front of a smaller one, lasts 10 hours.

Novas and Supernovas

- Location: Novas are largely confined to the Milky Way; supernovas can appear anywhere in the sky
- Visibility: A supernova can outshine all the stars in a galaxy
- Magnitude: Novas peak around magnitude 4 and on rare occasions magnitude 1; most supernovas peak below magnitude 14

A nova is a stellar outburst that occurs when a dense white dwarf gravitationally drains enough gas from a giant companion star to trigger a thermonuclear flash point. Within a week or two, a nova can erupt in brightness by 7 to 16 magnitudes, becoming as bright as magnitude 4. On rare occasions, a nova may exceed magnitude 1. Thereafter, it slowly fades over years or decades.

One of the brightest recent novas occurred in February 1992 in Cygnus. Its peak magnitude was 4.5, bright enough to be seen with the naked eye and an easy object in binoculars. Seventeen years before that another Cygnus nova topped out at magnitude 1.8.

Many amateur astronomers conduct nightly searches for novas using low-power telescopes or even 7 x 50 binoculars. They look for "intruder" stars shining among the familiar stellar patterns that they have committed to memory or that do not appear on their star charts.

A supernova is the cataclysmic death of a massive star. When a star goes supernova, it typically increases in brightness by 20 or more magnitudes, shining with the brilliance of several hundred billion suns. To hunt for supernovas, you need to scrutinize as many spiral galaxies as possible, night after night, looking for a "new," bright star within a galaxy's disk or outer arms. An essential piece of equipment is a telescope 8 inches (200 mm) or larger.

Supernovas are rare in our Galaxy—the last was seen in 1604—but dozens are seen each year in the myriad galaxies beyond.

The most recent naked-eye supernova was discovered on February 23, 1987 in the Large Magellanic Cloud, a nearby satellite galaxy to the Milky Way. That May, it reached a peak magnitude of 2.9 before rapidly fading below the naked-eye limit.

Cygnus Loop
The Cygnus Loop, or Veil Nebula, is the expanding shell of gas from a supernova explosion.

Supernova 1987A
A new star appeared within the Large Magellanic Cloud in February 1987. Seen here at lower right (with the Tarantula Nebula on the left), SN1987A was the first naked-eye supernova since 1604. Rings of light spread from the central star in this composite Hubble image (below).

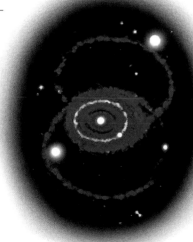

The Pleiades

- Constellation: Taurus
- Visibility: Northern Hemisphere and north of 59°S
- Magnitude: 1.2
- Apparent size: 1.8 degrees
- Distance: 400 light-years

M 45, the Pleiades star cluster, has been a source of wonder for centuries. To the naked eye, the individual members are not particularly bright. But as a collective, and sprinkled over an area of sky three times that of the Full Moon, they form a distinctive and beautiful nexus of stars.

Under good conditions, keen-sighted observers can make out seven stars—the cluster is often referred to as the Seven Sisters—although typically only six stars can be seen with the unaided eye. According to ancient lore, Merope, one of the sisters, married a mere mortal and for this reason shines less brightly than her siblings. The brightest member, Alcyone (Eta [η] Tauri), lies at the easternmost corner of the cluster.

With low magnification, an 8 inch (200 mm) telescope under a dark sky reveals that the four brightest stars shine with a kind of hazy quality, as if glowing behind slightly fogged glass. This is due to the interstellar dust cloud in which the cluster is embedded.

On an astronomical time scale, the Pleiades is a comparatively young cluster—probably no older than a few hundred million years. The light we see arriving tonight from this misty group of stars, however, is far younger. It began its journey about four centuries ago. Today, astronomers study the Pleiades to better understand the evolution of stars. Because of the cluster's rich history, and what astronomers are still learning about its stars, the Pleiades continues to be a source of great fascination.

Icy Glow
In a telescope at low magnificatrion, the Pleiades glow icy blue-white.

Bathed in Blue
Bathed in a blue reflection nebula, the brightest stars of the Pleiades make a spectacular close-up photo. Astronomers estimate the cluster contains 500 stars scattered across 100 light-years of space.

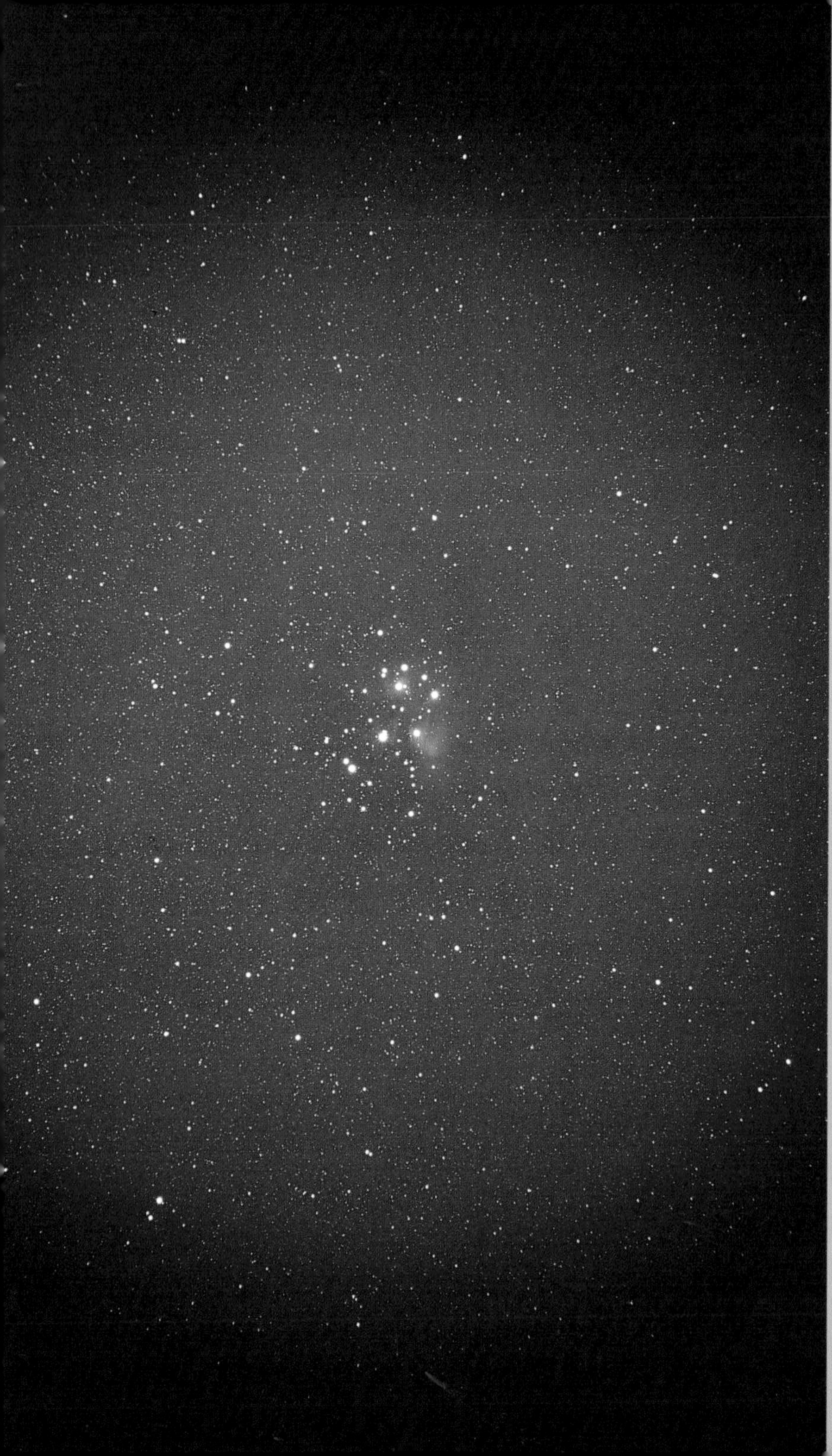

The Double Cluster

- Constellation: Perseus
- Visibility: Northern Hemisphere and north of 59°S
- Magnitude: 4.3
- Apparent size: 30 arcminutes each cluster
- Distance: 7,000 light-years (NGC 869); 8,100 light-years (NGC 884)

On a dark night in September, northern observers can see an elongated patch of hazy light between the constellations Perseus and Cassiopeia in the northeastern sky. If you turn binoculars on the region you will see a "figure-eight" of glittering stars; a 4 inch (100 mm) telescope at low magnification reveals two stellar aggregates side by side and equally bright. This is the Double Cluster—NGC 869 and NGC 884—and it is without doubt one of the most striking deep-sky wonders of the northern sky.

A 6 inch (150 mm) telescope at low magnifications shows a field strewn with glittering stars. NGC 869, the more westerly of the pair, is slightly brighter than 884, but both have highly concentrated centers and populations of about 700 stars. The edges of the clusters are indistinct, blending into the already rich starry background of the region.

Though they may look adjacent to one another in the sky, they actually lie at different distances. The distance to NGC 869 is thought to be about 7,000 light-years and for NGC 884, 8,100 light-years. Astronomers believe that they evolved from the same cloud of gas, however.

Both are young clusters, on the order of 10 million years, with NGC 884 thought to be slightly older. This is apparent in the number of evolved red supergiant stars scattered within it and toward its outer edges. With a telescope, you can see a few of these, particularly between the two clusters. These are massive stars that, within a few million years time, have rapidly evolved into old age.

Star Field
The Double Cluster (lower center) lies in one of the richest areas of the northern winter Milky Way. Even a casual sweep with 7 x 50 binoculars will bring into view numerous knots and clusters of stars.

Cluster Duo
Although they appear very close to one another in the sky, NGC 869 (right) and NGC 884 (left) are not really associated.

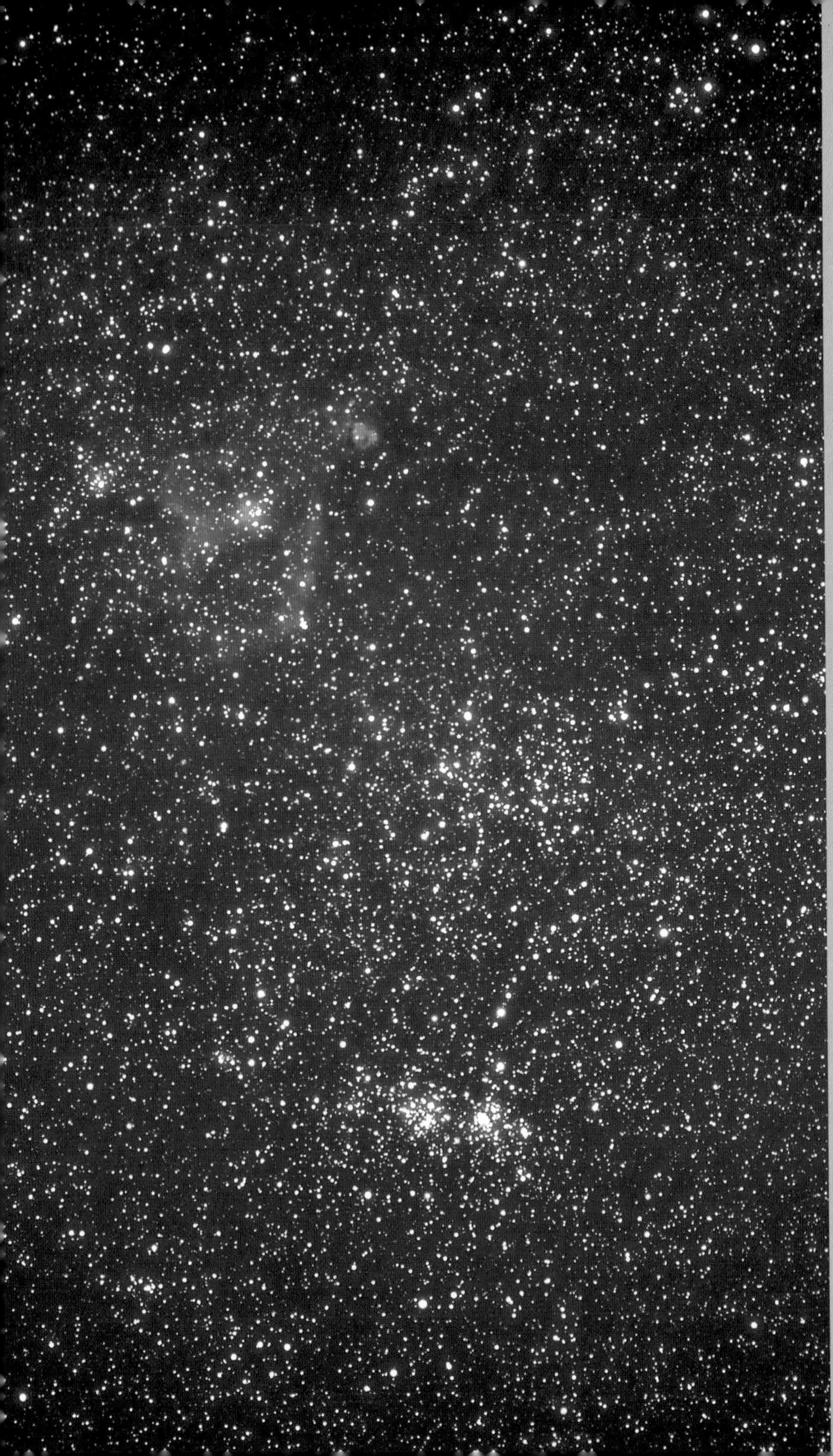

The Jewel Box Cluster

- Constellation: Crux
- Visibility: Southern Hemisphere and south of 23°N
- Magnitude: 4.2
- Apparent size: 10 arcminutes
- Distance: 8,800 light-years

Lying just south of the bright star Mimosa (Beta [β] Crucis) in Crux is a coarse aggregate of about 50 blue and red stars that on the whole resembles a flashy necklace of rubies and sapphires. This bespangled object is the Jewel Box (NGC 4755), one of the loveliest open clusters in the southern sky. Like the Double Cluster (see p. 368), it is relatively young, at about 10 million years, but lies at a greater distance, about 8,800 light-years.

The stars of the Jewel Box range in magnitude from 6 to 10. The brightest members, including 6th-magnitude Kappa (κ) Crucis, form a wedge-shaped pattern confined to an area of sky smaller than the apparent diameter of the Full Moon. At the center of the cluster a brilliant 7th-magnitude red supergiant provides a startling color contrast to the surrounding blue and white stars. The bulk of the fainter members lie south and southeast of the cluster's center.

The Jewel Box can be seen with the unaided eye in a dark sky—it resembles a bright "knot"—but is best observed using either binoculars or a low-power telescope. With a telescope, keep in mind that too much magnification will "uncluster" the cluster. It can be most conveniently observed when it is highest in the sky, on early June evenings.

Stargazers should also turn their attention to the remarkable region surrounding the Jewel Box. Just south lies the starless void of the Coalsack, a cloud of dense interstellar dust and one of the nearest of the dark nebulas. This, too, is apparent to the naked eye. Binoculars also reveal many dark lanes and dust filaments in and around the Jewel Box, the raw material of stars yet to be.

Color Contrasts
A wonderful feature of the Jewel Box is the contrasting colors of the line of three stars at its heart. The English astronomer John Herschel gave the cluster its name in the 1830s, comparing it to a "casket of variously coloured precious stones."

Gems in Crux
The Jewel Box looks like a small star, tucked under Beta Crucis, which forms the bright lefthand point of Crux. Just beneath it is the dark mass of the Coalsack.

371

The Hercules Cluster

- Constellation: Hercules
- Visibility: Northern Hemisphere and north of 47°S
- Magnitude: 5.9
- Apparent size: 17 arcminutes
- Distance: 21,000 light-years

A favorite among northern deep-sky observers is the Hercules Cluster, or M 13, an effervescent globe of stars. This globular cluster lies poised a little more than 2 degrees south of Eta (η) Herculis, which is the northernmost star in the "keystone" asterism of the constellation. On dark, clear nights away from city lights, you can see the Hercules Cluster with the naked eye, looking like a faint, fuzzy star. Binoculars show it as a decidedly woolly patch of light that cries out for closer scrutiny with a telescope.

In small telescopes, M 13 looks like the nucleus of a comet that is beginning to brighten. A 4 inch (100 mm) scope, however, brings out this object's stellar nature. The center is a homogenous glow, but you can just make out stars around the edges. In a telescope of 10 to 13 inches (250 to 330 mm), and at magnifications of 30x or 40x, the Hercules Cluster is glorious. The core scintillates with stars, and star chains radiate outward in all directions.

This is a particularly rich and compact cluster. Astronomers estimate that it contains at least a million stars within a region only 100 light-years across. That works out to about one star per cubic light-year near the center. The night-sky view from a hypothetical planet near the heart of the cluster would be filled with a dazzling array of stars. There would never be a truly dark night on such a world, nor a universe beyond to appreciate.

The Hercules Cluster is ancient. In fact, at about 12 billion years of age it is one of the oldest things that you will ever see.

Comet or Cluster?
Like all globulars, the Hercules Cluster has a cometlike appearance in a small telescope. (Novice comet-hunters often mistake globulars for comets.) It appears high overhead in Northern Hemisphere skies toward the end of July.

City of Stars
A medium-size telescope aimed at a dark sky will reveal a beautiful granular sphere with strands of stars radiating out from its center. The Hercules Cluster's stars are all fainter than magnitude 11.

Omega Centauri

- Constellation: Centaurus
- Visibility: Southern Hemisphere and south of 35°N
- Magnitude: 3.7
- Apparent size: 36 arcminutes
- Distance: 17,000 light-years

One of the most magnificent objects visible in small telescopes is the great globular cluster Omega (ϖ) Centauri (NGC 5139). It can be seen with the naked eye as a "woolly" star, but in binoculars it has a distinctly granular appearance. Through a 4 inch (100 mm) telescope, with medium magnification, the object resolves into a great orb of stars.

Being located at the southerly declination of −47 degrees, Omega Centauri is not often observed in the Northern Hemisphere. Still, it can be seen low in the southern sky from latitude 35°N in late April and early May when it crosses the celestial meridian. From latitude 30°N, its greatest altitude is slightly over 12 degrees. Look 36 degrees due south of Spica, brightest star in the constellation Virgo. This is roughly equal to the distance covered by three fists held at arm's length.

Visually, the cluster has a magnitude of about 4 and an apparent diameter of half a degree, the same as the Full Moon. This translates into a true diameter (at its distance of 17,000 light-years) of 150 light-years. It is probably larger, perhaps as much as 300 light-years. The star-rich core of the cluster is about 100 light-years across.

Telescopically, Omega Centauri is spectacular. The flattened, seething sphere of stars gradually gives way on the edges to a diffuse glitter of thousands of diamond-like chips that trail out into the dark of space. In a large telescope, this object is simply extraordinary, showing stellar points all the way across the center.

Grand Globular
Omega Centauri is a spectacular sight in any telescope over about 6 inches (150 mm).

Stellar Oval
Unlike most globulars, Omega Centauri is oval rather than round in shape. This distinctive appearance comes from the movement of its stars, of which there are roughly a million.

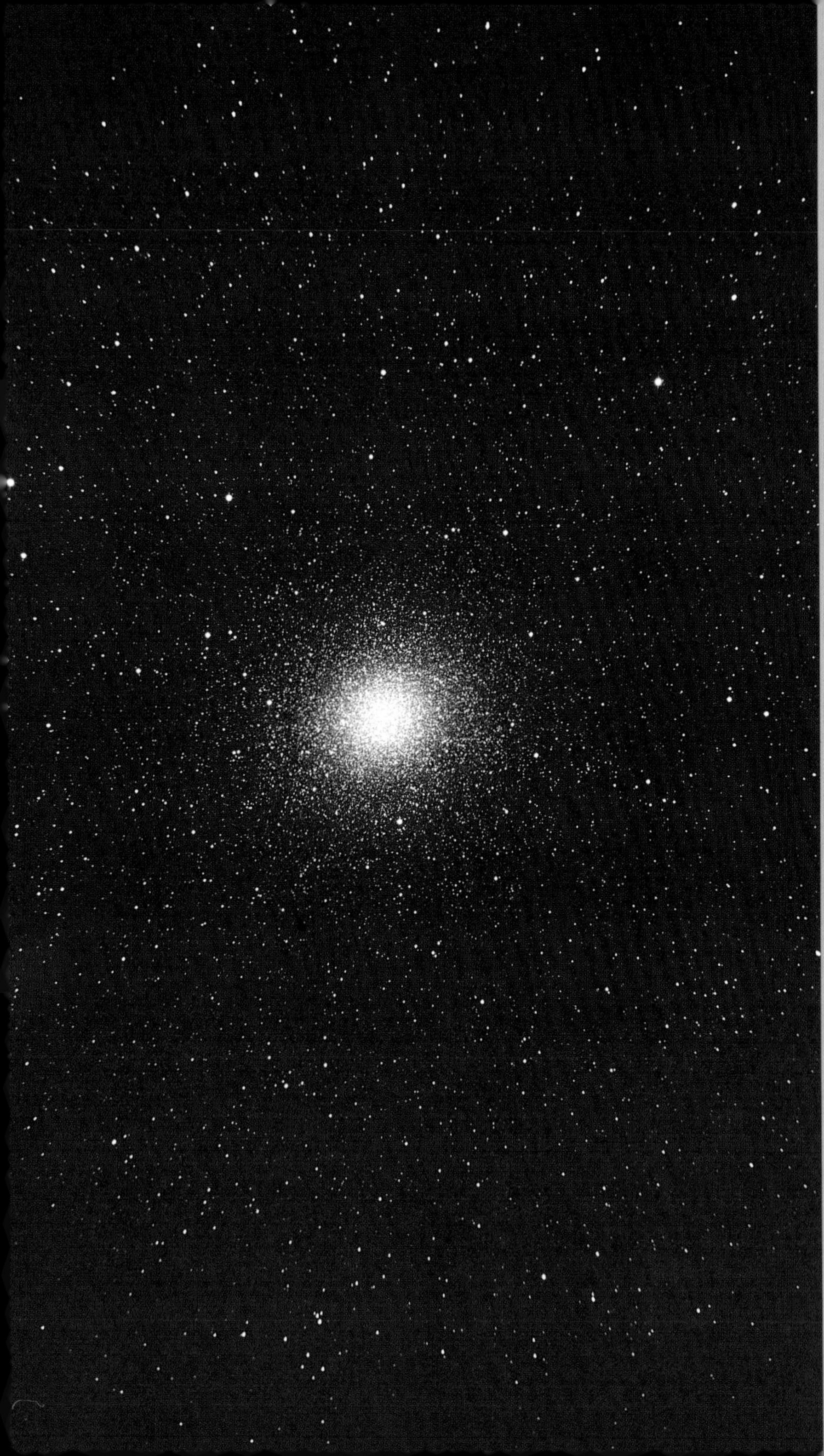

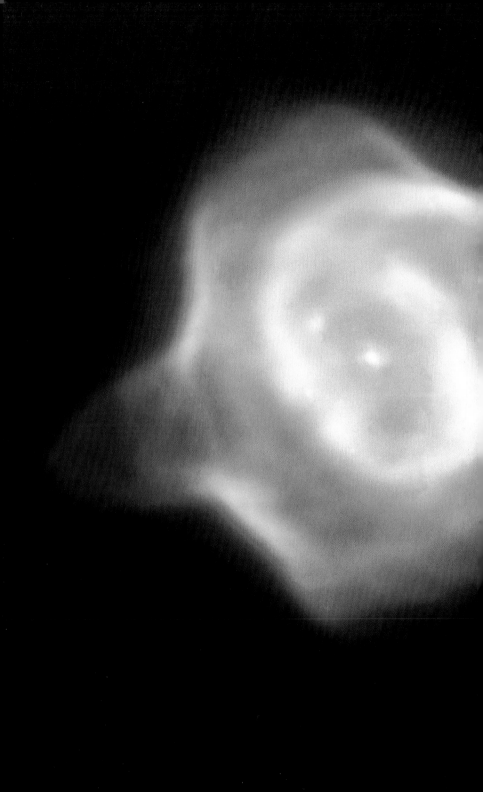

Observing
Nebulas

Of all the sights in the night sky,
the delicate clouds of gas and dust
known as nebulas—the birthplaces
and graveyards of the stars—are
among the most stunning.

"A thousand little wisps, faint nebulae,
Luminous fans and milky streaks of fire…"

from *The Torch-Bearers*,
Alfred Noyes, English poet, 1880–1958

The Owl Nebula

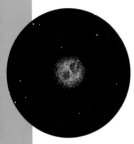

- Constellation: Ursa Major
- Visibility: Northern Hemisphere and north of 28°S
- Magnitude: 9.9
- Apparent size: 2.8 arcminutes
- Distance: 1,300 to 2,300 light-years

The Owl Nebula (M 97) has a reputation for being one of the most difficult objects in the Messier Catalog to find with a small telescope, but it should present no problems as long as you have at least a 3 inch (75 mm) scope, a low-power eyepiece and, most important, dark, clear skies.

Owl's Eyes
Some 2 to 3 light-years in diameter, the Owl Nebula is one of the sky's largest planetaries. But because of its low surface brightness, you need a medium-size telescope to gain even a hint of the owl's dark eyes, as seen here.

In the Limelight
In this image, the Owl Nebula (upper right) shares the limelight with the spiral galaxy M 108 (lower left).

To locate it, look just below the front of the Big Dipper's cup; more specifically, it is about 2 degrees southeast of Merak, the southernmost star in the lip of the cup. You should be able to see a gray, featureless oval. Higher magnification reveals little in the way of detail, though a 6 or 8 inch (150 or 200 mm) telescope may bring out a suggestion of the owl's dark "eyes."

A nebular filter helps enhance the contrast between the object and the background sky, but if you do not have one of those, you can employ averted vision (see How to See the Most, p. 111). Basically, by shifting your gaze slightly away from the nebula, you stimulate the more light-sensitive peripheral cells in your eye. The edges between the nebula and the dark sky suddenly become easier to see.

The Owl Nebula is formed from the expelled shell of a dying star. At the center of the nebula lies the star itself, but at magnitude 14, it is difficult to see except in telescopes over 10 inches (250 mm) in diameter. The dark perforations in the shell, which may be glimpsed in telescopes with apertures of 12 inches (300 mm) or more, are likely the result of turbulence within the expanding gas. There is also a very faint circular outer ring enclosing the owl's face, but this only appears in long-exposure photographs.

The Trifid Nebula

- Constellation: Sagittarius
- Visibility: Southern Hemisphere and south of 60°N
- Magnitude: 5
- Apparent size: 29 x 27 arcminutes
- Distance: uncertain, ranging from 2,200 to 6,000 light-years

The Trifid (M 20) is one of several major gaseous nebulas in Sagittarius. Located about 6 degrees northwest of Kaus Borealis (Lambda [λ] Sagittarii), 7 x 50 binoculars show a small hazy patch encompassing a 6th-magnitude star. The nebula lies just 1.2 degrees north of the brighter Lagoon Nebula (see p. 382). You will need a 4 to 6 inch (100 to 150 mm) telescope to show the nebula's most distinctive characteristic, which also gives it its name—three irregularly shaped lobes of gas trisected by dark lanes of dust. These rifts, however, have a much more dramatic appearance in telescopes of 8 inches (200 mm) and larger, rendering the nebula as a bloomed-out rose.

The central star, cataloged HN 40, is actually a six-star system located on the inside apex of one of the dark lanes. Small telescopes show this star as a double, while all six can be detected in larger instruments. The primary star is a young, hot, and very luminous supergiant. Long-exposure photographs reveal that the Trifid is shaded in hot pink and blue hues. The hot pink is hydrogen gas whose atoms have been excited, or ionized, by ultraviolet radiation from the supergiant, while the blue tinge is caused by starlight reflecting off the dust lying just north of the main nebular mass. Thus the Trifid is both an emission and reflection nebula.

Distance estimates to the Trifid Nebula range from 2,200 to more than 6,000 light-years. Based on the higher distance value, astronomers believe the Trifid is not associated with the neighboring Lagoon Nebula.

Color Correction
Long-exposure photos such as these show the emission and reflection parts of the Trifid Nebula as red and blue, respectively. Do not expect to see anything like this in a telescope, however. Our eyes are just not sensisitive enough to detect these colors, and in the eyepiece, the Trifid looks gray or gray-green, as do all nebulas.

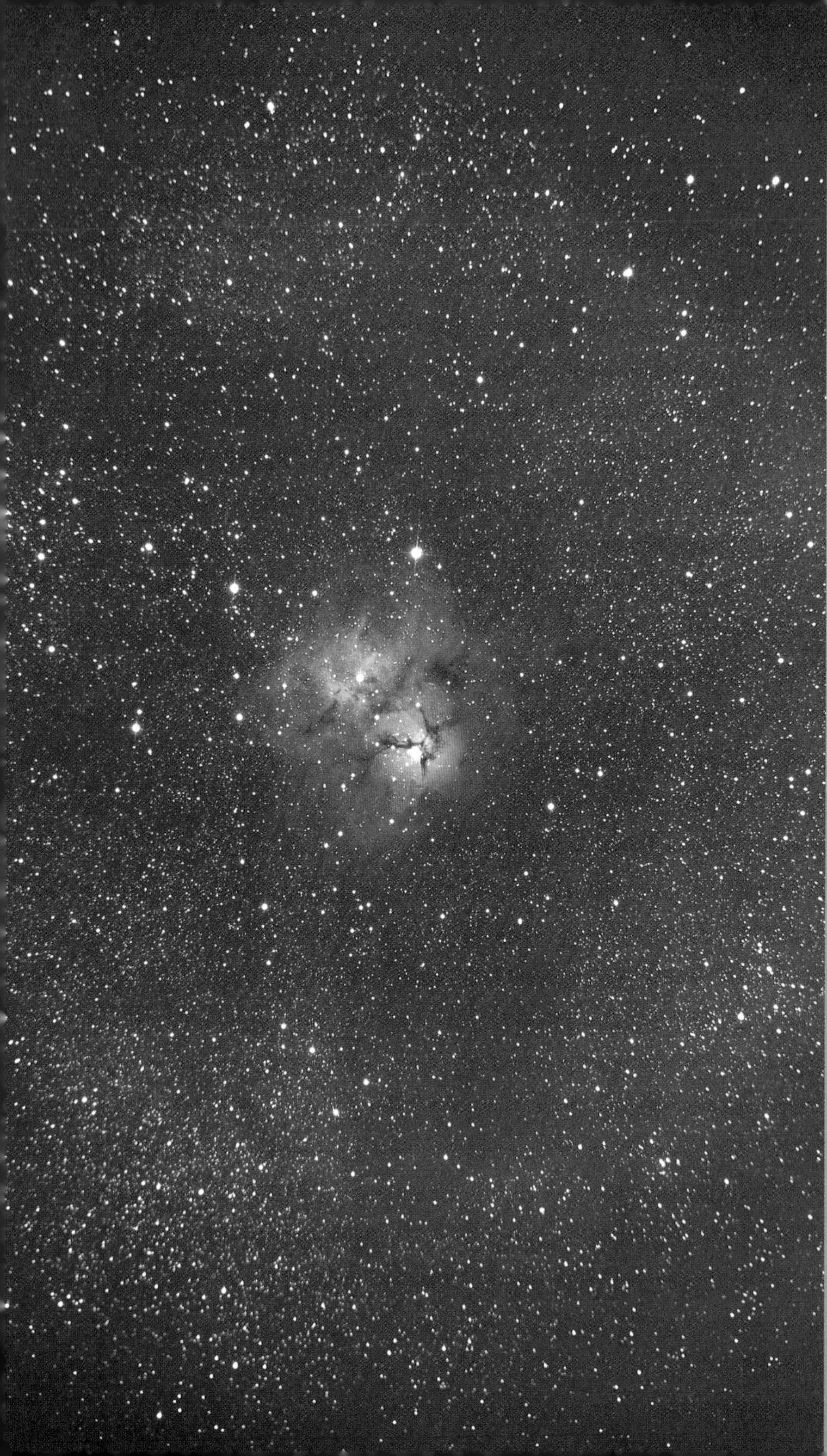

The Lagoon Nebula

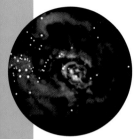

- Constellation: Sagittarius
- Visibility: Southern Hemisphere and south of 60°N
- Magnitude: 5.8
- Apparent size: 90 x 40 arcminutes
- Distance: 4,500 light-years

The Lagoon Nebula (M 8) is not only one of the most striking deep-sky objects in Sagittarius, but in the entire night sky, ranking near the Orion Nebula (see p. 394) in beauty and brightness. It can easily be seen in binoculars as a soft smear of light with embedded stars 6 degrees north of Alnasl (Gamma [γ] Sagittarii). Closer inspection with a telescope brings out a complex, textured nebula, sprinkled throughout with tiny stars, especially in its southwest quadrant. It is divided almost in half by a curving dark channel of obscuring dust—the "lagoon"— that is interspersed here and there with filaments of glowing gas. A nebular filter brings out this feature particularly well.

The nebula was first recorded by the English astronomer John Flamsteed around 1680, and in 1764 it was logged in Charles Messier's list of nebular objects. Messier noted that the object had an angular diameter of 30 arcminutes (the apparent diameter of the Full Moon). Indeed, to see this object in its entirety, you will need a wide-field eyepiece for a telescope, or 7 x 50 or larger binoculars.

Details seen in 4 to 8 inch (100 to 200 mm) telescopes betray the true nature of the Lagoon Nebula. Like the Trifid and Orion nebulas, it is a star-forming region. Two bright stars only 3 arcminutes apart dominate the western half of the nebula. The southern component is 9 Sagittarii, thought to be the primary source of the nebula's fluorescence. The brightest portion of the nebula is in this same region. The eastern half contains a loose star cluster, NGC 6530—these are stars that have coalesced from the nebula itself.

Exploring Sagittarius
Sagittarius is dotted with impressive star clouds, nebulas, and star clusters, making it a joy to explore by telescope or binoculars. The Lagoon Nebula is the brightest and largest nebula in the region. In this image, it stretches upward toward NGC 6559, marked by blue, ball-like structures. South is to the left.

Seeing the Lagoon
Even in a small telescope, a dark channel, or lagoon, can be readily seen winding its way between the brightest regions of the nebula. A nebular filter accentuates the view.

The Eagle Nebula

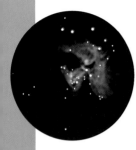

- Constellation: Serpens
- Visibility: Southern Hemisphere and south of 71°N
- Magnitude: 6
- Apparent size: 34 x 27 arcminutes
- Distance: 7,000 light-years

The Eagle Nebula (M 16) is a combination nebula and star cluster lying a little over 2.3 degrees west-northwest of Gamma (γ) Scuti in Serpens. Like the Lagoon Nebula (see p. 382), this is a stellar nursery. Binoculars show a compact cluster of about a dozen bluish stars, but in telescopes of 8 inches (200 mm) or more a faint haze can be seen enshrouding them. Averted vision (see How to See the Most, p. 111) really helps bring out the contrast in this nebula.

Further scrutiny at higher magnifications reveals a dark, curtained area in the northern portion that narrows to a cone toward the nebula's center. Nebulosity stretches away to the east and west in two mottled, cauliflower-like lobes. A large telescope unveils a dark blot punctuating the northeastern edge of the nebula, which is caused by intervening dust. The nebula lies at a distance of about 7,000 light-years, meaning the visible portion is some 70 light-years across.

The cluster of about 50 stars occupies a northwestern spur that projects outward from between the lobes. These are mostly hot supergiants that have spawned from the dense, dusty regions of the nebula. They, like the hot stars in the Lagoon and Trifid nebulas, are responsible for exciting the gas to incandescence.

Long-exposure photographs of this object reveal its true glory, for at the center lies a dark mass of dust that looks like a great hand reaching toward the northwest. In April 1995, the Hubble Space Telescope zeroed in on this part of the nebula revealing dense, light-year-long fingers of dust being blasted by radiation from nearby hot stars, exposing newly born stars (see A Look Inside the Eagle Nebula, p. 186).

A Dazzling Eagle
The Eagle Nebula is a dazzling combination of nebula and star cluster, adorned by lanes of dark dust. The nebula lies in a particularly rich part of the Milky Way, littered with regions of actively forming stars. The average age of the nebula's stars is less than a million years.

Dust Lanes
Dark dust lanes run through the white heart of the Eagle Nebula.

The Omega Nebula

- Constellation: Sagittarius
- Visibility: Southern Hemisphere and south of 68°N
- Magnitude: 7
- Apparent size: 20 x 15 arcminutes
- Distance: 3,000 light-years

Just 2.2 degrees south of the Eagle Nebula is another fine object for small telescopes, M 17, the Omega Nebula. The brightest portion of this nebula is an 8-arcminute-long east-to-west bar that is just visible in 7 x 50 binoculars. Small telescopes in dark skies bring out an arch of nebulosity extending to the south from the western end of the bar. The English astronomer William Herschel interpreted this arch as the curve in the Greek capital letter omega (Ω), but it has also been envisaged as the head of a swan and a horseshoe. The hollow within the arch is the result of obscuring dust. Overall, the nebula looks like a number 2 with a lengthened foot.

Glowing in the Dark
Hidden from view by dust and gas, young, hot stars make the Omega Nebula glow like a neon sign. Astronomers think that star formation is either still active in the nebula, or stopped relatively recently.

Dark dust also has a hand in giving the Omega Nebula a contrasty appearance to the north and northwest, making these borders very prominent even in small telescopes, particularly when using averted vision (see How to See the Most, p. 111). Filaments of dust crisscross the bar, giving it a patchy appearance that can be seen in telescopes of 6 inches (150 mm) or more. Faint, tenuous components of the nebula swirl off the eastern end of the bar and appear to loop over and around the entire complex.

Although the Omega Nebula does not contain a prominent star cluster, the field is sprinkled with about three dozen or so stars of magnitude 9 and fainter. Astronomers believe that the stars responsible for illuminating the nebula are embedded in or hidden by dust and gas. At least 35 stars are definitely associated with the nebula, comprising a total mass some 800 times that of the Sun.

Omega and Trifid
The Omega Nebula lies at the top of this view of part of the Milky Way, with the Trifid Nebula at center bottom. Surrounding dark dust makes the Omega a high-contrast object for small telescopes.

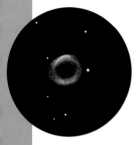

The Ring Nebula

- Constellation: Lyra
- Visibility: Northern Hemisphere and north of 50°S
- Magnitude: 9.0
- Apparent size: 1.2 arcminutes
- Distance: between 2,000 and 5,000 light-years

Framed neatly between the stars Sulafat and Sheliak—Beta (β) and Gamma (γ) Lyrae, respectively—is the Ring Nebula (M 57), one of the most popular planetary nebulas with amateur astronomers. You will need at least a 3 or 4 inch (75 or 100 mm) telescope, moderate magnification, and dark skies to see the hole in the "smoke ring." A 6 inch (150 mm) scope shows the ring as a glaucous, slightly elongated orb with a dark gray void in the middle.

Being only a little more than 1 arcminute in apparent diameter and so well defined, the Ring Nebula can withstand high magnifications. Even at 150x, the oval outline remains sharp and the hole is veiled with a thin filmy haze. At higher magnifications and under exceptional seeing conditions the oval appears to have pointed ends. In larger telescopes, the nebula's characteristic blue-green color (caused by ionized oxygen) begins to emerge.

The Ring Nebula is between 2,000 and 5,000 light-years from Earth and has a diameter of about 1 light-year. It is the exhaled shell of a dying star. The faint central star (magnitude 15.4) was once larger than the Sun. Now near the end of its life, the star's inner core has collapsed on itself and the resulting

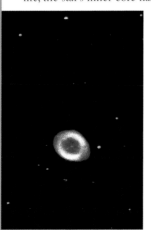

internal heat has puffed the star's atmosphere into space. The remnant is now a tiny white dwarf, about the size of Earth. Unfortunately, the central star is visible only in telescopes of 13 inch (330 mm) caliber or so, and even then with difficulty. Skies must be very dark, clear, and steady.

Hubble images show that what looks like a ring is actually barrel-shaped. So, rather than looking into a ring, we are looking down into a cylinder seen almost end-on.

A Ring Lit by a Dwarf
The Ring Nebula's central star, a white dwarf, is very hot and produces intense ultraviolet radiation that results in the bright fluorescence of rarified gases in the nebula. The ring is about 1 light-year wide and is still expanding.

First Description
Backyard astronomers looking for M 57 might like to keep in mind the initial description of this object, by the Frenchman Antoine Darquier de Pellepoix in 1779: "a dull nebula, but perfectly outlined; as large as Jupiter and looks like a fading planet."

Celestial Doughnut
For obvious reasons, the Ring Nebula is often referred to as the "smoke ring" or "doughnut" of the night sky.

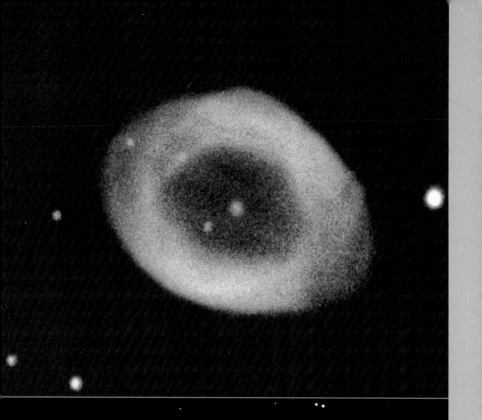

The Dumbbell Nebula

- Constellation: Vulpecula
- Visibility: Northern Hemisphere and north of 51°S
- Magnitude: 7.3
- Apparent size: 8 x 4 arcminutes
- Distance: 1,000 light-years

Like the Ring Nebula, the Dumbbell Nebula (M 27) is in the class of planetary nebulas—essentially the outer shell of a dying star. This phase of the death of a star does not last very long, however, which is why there are not that many planetaries plotted in the sky. After about 50,000 years, they simply dissipate into space.

The Dumbbell Nebula is large, bright, and easy to locate. Using a telescope or binoculars, carefully scan the sky 3 degrees north of Gamma (γ) in the constellation Vulpecula. Look for a round disk of gray light. (In binoculars, this object will appear as a blurred dot.)

You need dark skies and a telescope with an aperture of at least 4 inches (100 mm) to see any detail. Use a medium-power eyepiece and try using the technique of averted vision (see How to See the Most, p. 111). As you increase magnification, the nebula's edge takes on a greater presence as contrast increases. The smoky disk fills the field of view like an airbrushed cloud against the dark sky, its oval center oriented northeast to southwest. You can probably go as high as 150x in a 6 inch (150 mm). Higher magnifications, however, will likely require greater

Closing In
The Dumbbell Nebula is a diaphanous smudge in space when seen in a 5 inch (125 mm) telescope (top). The larger the instrument that is used, the greater the details revealed in the nebula's disk. In the bottom photo, note the area of faintly glowing gas on the nebula's periphery.

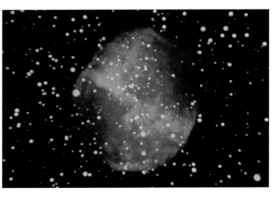

aperture because the object just spreads out and darkens at larger image scales. The central star requires at least a 13 inch (330 mm) telescope. Overall, the Dumbbell looks more like an apple core or hourglass than a weightlifting apparatus.

The distance to the Dumbbell is thought to be about 1,000 light-years, which would give this object a true diameter of between 2 and 3 light-years. That makes it one of the largest of the known planetary nebulas.

Stars in the Background
A special imaging process was used to highlight the Dumbbell's background of faint stars. An obvious consequence of this process is a shift in colors toward the blue-green.

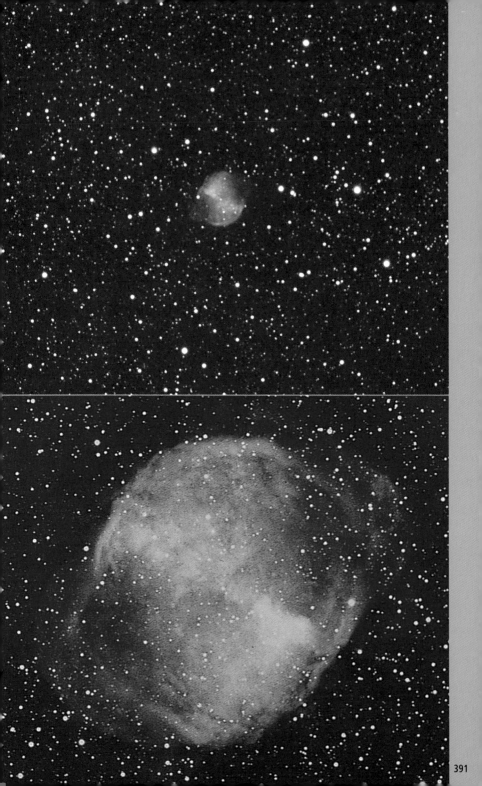

The Crab Nebula

- Constellation: Taurus
- Visibility: Northern Hemisphere and north of 62°S
- Magnitude: 8.4
- Apparent size: 6 x 4 arcminutes
- Distance: 4,000 light-years

Taurus rises early in northern winter evenings, with its long horns trailing to the northeast. Just slightly over 1 degree northwest of Zeta (ζ) Tauri—the star marking the tip of the southeastern horn—lies the Crab Nebula, number one in Charles Messier's catalog. This intriguing object is the gaseous remains of a supernova recorded by Chinese and Japanese astronomers in AD 1054.

This is one of the few supernova remnants that can be detected by binoculars—but only if your sky is dark enough. With angular measurements of only 6 x 4 arcminutes, it appears as a small, woolly star in a pair of 7 x 50s. A 4 to 8 inch (100 to 200 mm) telescope at medium magnification shows a wispy oval without much texture. Unfortunately, increased magnification does little to improve detail. Larger instruments, however, reveal serrated edges and filaments along the outer edges of the nebula. Long-exposure photographs and digital images reveal a beautiful, tortured object crisscrossed by tendrils of hot gas that have been swept up into shockwaves from the supernova blast.

At the heart of the Crab lies the core of the exploded star itself, still spinning from the blast observed from Earth. At magnitude 16, however, it can be seen only in very large telescopes. The star is composed almost entirely of compressed neutrons and has a diameter of about 6 miles (10 km). Even more astonishing, this neutron star beams concentrated radio waves in Earth's direction with every rotation, like a lighthouse—30 times a second. Such an object is called a pulsar.

From M 1 to Crab
Photos show that M 1 bears little resemblance to a crustacean. The name was applied when the Earl of Rosse, a 19th-century English astronomer, made a sketch of M 1 based on observations made with his giant telescope. The sketch looked a little like a crab—and a nickname was born.

Twisted Look
The twisting filaments of the Crab Nebula are caused by hot gas flung out when the original star went supernova. In this image, the colors have been enhanced to emphasize the nebula's structure. The filaments show as greens, reds, and yellows; the bluish inner glow indicates an intense magnetic field.

Hazy View
In a 4 inch (100 mm) telescope, the Crab looks like a small patch of haze. To see the nebula's distinctive tendrils, you will need a scope larger than 8 inches (200 mm).

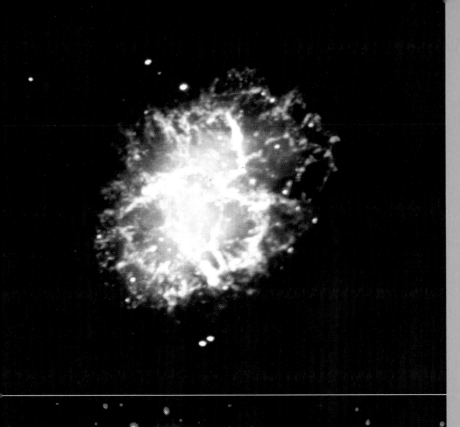

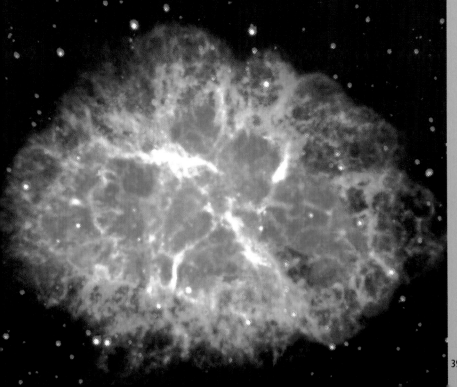

The Orion Nebula

- Constellation: Orion
- Visibility: Northern and Southern hemispheres
- Magnitude: 4
- Apparent size: 65 x 60 arcminutes
- Distance: 1,500 light-years

One of the most stunning sights in the heavens is the Orion Nebula (M 42), also known as the Great Nebula. To the naked eye, it appears as a dim patch of haze in the middle of the Hunter's sword. Binoculars easily show this object to be nebulous in nature, but only in a telescope can its beauty be truly appreciated.

A 6 or 8 inch (150 or 200 mm) telescope and magnifications of 40x or 50x reveal a billowy, fan-shaped cloud shot through with glittery stars. The eastern edge of the fan is brightest and therefore exhibits the greatest contrast. Shock waves from the solar winds generated by hot, young stars at the heart of the nebula have piled up gas and dust along this region. In contrast, the western edge is dappled with swirls of glowing gas and dust and is well worth exploring at higher magnifications. In larger telescopes, the Orion blossoms into a complex, nearly three-dimensional region of light and shadow. The outer fringes take on a petal-like appearance, and can be traced out as far as half a degree from the nebula's center.

At the nebula's heart lies a square knot of four bright stars called the Trapezium—a multiple system of newborn stars. The Trapezium is actually part of a larger collection of young stars and protostars called the Orion association, which contains over a thousand stars. The stars of the Trapezium are some of the association's oldest members, and yet they date back only about a million years.

Just to the north of the main body of the Orion Nebula is M 43, a small patch of nebulosity.

Orion's Jewels
The constellation Orion is home to two celestial wonders—the Orion Nebula (right) and the Horsehead Nebula (left).

Impressive Sight
The Orion Nebula is one of the most impressive sights for skywatchers. It is pictured in the upper image with its companion, M 43, together with NGC 1977, a blue nebula. The lower, specially processed image is detailed enough to reveal the stars that illuminate both M 42 and M 43.

Observing
Galaxies

Without a doubt, galaxies can be
challenging to observe, but what
other endeavor extends your vision
across such great gulfs of space?
The quest to behold the diverse and
enigmatic galaxies, each shining with
the combined light of billions of
stars, is deep-sky observing at its
most profound.

*"Eventually, we reach the...utmost limits
of our telescopes. There, we measure shadows,
and we search among ghostly errors of
measurement for landmarks that are
scarcely more substantial."*

Edwin Hubble
U.S. astronomer, 1889–1953

M 81

- Constellation: Ursa Major
- Visibility: Northern Hemisphere and north of 15°S
- Magnitude: 6.8
- Apparent size: 16 x 10 arcminutes
- Distance: 11 million light-years

Located 10 degrees northeast of the cup of the Big Dipper, M 81 is one of the finest examples of a symmetrical spiral galaxy in the sky. At magnitude 7, it is conspicuous in jet-black skies and can also be seen in 7 x 50 binoculars.

M 81 is oriented nearly 45 degrees from being face-on, and appears as a distinct oval disk with a bright, dense core. In dark, lucid skies an 8 inch (200 mm) telescope at a magnification of 40x even shows traces of two tenuous arms spiraling outward from the bright portion of the disk. The spiral arm extensions that trail farthest from the disk, however, are beyond the light-gathering power of even large telescopes.

The apparent dimensions of M 81 are 16 by 10 arcminutes, with the long axis of the oval oriented north to south. Detailed images reveal an even larger disk, some 26 arcminutes long, nearly the apparent size of the Full Moon.

Just half a degree north of M 81, and in the same low-power field of view, is the unusual galaxy M 82, which is seen edge-on. Try sweeping across both galaxies to compare their shapes.

M 82 is classified as a starburst galaxy. Tens of millions of years ago, M 81 and M 82 passed very near each other. The gravitational interaction greatly affected both systems, spawning new stars in the central region of M 81 and a burst of star formation in M 82.

Together, M 81 and M 82 form the nucleus of a small group of galaxies that is one of the nearest to our Local Group, only 11 million light-years away.

Classic Spiral
M 81 is one of the most symmetrical spiral galaxies in the heavens. The large, bright core contains most of the galaxy's 250 billion stars. In 1993, a supernova occurred in M 81 and was discovered by an amateur astronomer.

Outside View
Astronomers say that M 81 is probably a fair representation of how the Milky Way Galaxy would look from the outside.

M 100

- Constellation: Coma Berenices
- Visibility: Northern Hemisphere and north of 68°S
- Magnitude: 9.4
- Apparent size: 7 x 6 arcminutes
- Distance: 56 million light-years

With a diameter of more than 120,000 light-years, M 100 is among the largest spiral galaxies in the Virgo cluster of galaxies. Unfortunately, its colossal size does not make it easy to observe. Located in the constellation Coma Berenices, the galaxy can be seen through a small telescope, but aside from its bright, egg-shaped core, it is practically featureless in anything smaller than an 8 inch (200 mm) telescope.

M 100, with its estimated hundred billion Suns, lies 56 million light-years away (give or take 6 million), a distance astronomers determined by measuring the luminosity of a number of the galaxy's Cepheid variables. It is known as a grand-design galaxy because of its majestic spiral form. Images and photographs of M 100 reveal two prominent spiral arms, as well as an array of secondary ones and dust lanes that can be traced back toward the galaxy's nucleus. The arms are festooned with bright star-forming regions, some of which have been sites of past supernovas.

To find M 100, look 8 degrees east of Denebola, the magnitude 2 star that forms the tail of Leo. The nucleus is oriented not quite face-on from southeast to northwest. Telescopes of 8 and 10 inches (200 and 250 mm) easily show the galaxy's magnitude 9 nucleus. To detect hints of spiral structure, however, you need very dark skies, low magnification, and a little ingenuity. Once the galaxy is centered, use averted vision (see How to See the Most, p. 111) and then very gently tap the telescope tube. The slight jiggling "turns on" the eye's motion sensors, enabling you to detect the tenuous spiral arms just beyond the bright nucleus.

Seeing Spiral Structure
M 100 is a large face-on spiral galaxy glowing at magnitude 9.4. Its brilliant, starlike core makes it an easy object to locate, but do not expect to see a distinct spiral structure, as in photos such as this. In 8 inch (200 mm) telescopes or smaller, it can resemble a dim globular cluster. Ghostly arms emerge with the use of a larger scope under dark skies.

Largest in the Cluster
At 7 arcminutes across, M 100 presents the largest apparent size of any galaxy in the Virgo cluster.

M 87

- Constellation: Virgo
- Visibility: Northern Hemisphere and north of –71°S
- Magnitude: 9
- Apparent size: 7 arcminutes
- Distance: 70 million light-years

Located off the hindquarters of Leo near the center of the Virgo galaxy cluster, M 87 is one of the largest galaxies in the nearby universe. It is also a prodigious source of radio energy as well as a powerful X-ray source. If our eyes were sensitive to these wavelengths, this galaxy would light up the sky like the Full Moon.

Unlike spiral galaxies, in which the observer can coax out some tenuous spiral structure, M 87 is a giant elliptical galaxy, like the bulge of a spiral galaxy but without the arms. Its 7 arcminute disk, however, is quite bright at magnitude 9, appearing as an oval-shaped, cometlike object in 11 x 80 binoculars and a bright, diffuse sphere in a small telescope.

The source of this galaxy's intense energy is a supermassive black hole churning away at its core. The black hole occupies an area not much larger than our Solar System, but it is packed with the mass of five billion Suns.

During M 87's existence, it has swallowed and assimilated other galaxies, making it one of the galactic heavyweights of the known universe. A powerful beam of radio, X-ray, and visible light energy gushes from the galaxy's core at extraordinary velocities. In large telescopes, in fact, this jet can be seen pointing outward like the hour hand of a clock.

To find M 87, scan the region roughly between the stars Denebola in Leo and Epsilon (ε) Virginis, into the heart of the Virgo cluster. Do not be confused by the bright pair of galaxies M 84 and M 86. M 87 stands on its own a little over a degree east of these two.

A Giant in Virgo
Some 1,000 systems make up the Virgo cluster of galaxies. M 87 (at upper right) is one of the largest.

M 87's Retinue
M 87 is surrounded by a collection of over 4,000 globular star clusters, which appear here as fuzzy "stars." Two companion galaxies are also visible.

Cosmic Jet
Long-exposure photos, such as the main image, reveal many details of M 87 and its surrounds, but tend to "burn out" the center of the galaxy. A short exposure (inset) shows something very different—a jet of gas emanating from the nucleus. About 4,000 light-years long, the jet is thought to be a high-speed beam of particles, probably associated with a central black hole.

403

The Blackeye Galaxy

- Constellation: Coma Berenices
- Visibility: Northern Hemisphere and north of 62°S
- Magnitude: 8.5
- Apparent size: 9 x 5 arcminutes
- Distance: 12 million light-years

Lying snug within the faint L-shaped constellation of Coma Berenices is a unique galaxy with a unique name: the Blackeye Galaxy (M 64). The nickname arises from an unusually broad band of obscuring dust crossing the galaxy's midsection on its northeastern flank, giving it the appearance of a closed human eye with the proverbial shiner. The Blackeye Galaxy is located 10 degrees north of Epsilon (ε) Virginis, or 5 degrees northwest of 4th-magnitude Diadem (Alpha [α] Comae Berenices). Zooming in on this field, look for 5th-magnitude 35 Comae Berenices. The galaxy lies 1 degree to the east-northeast.

At magnifications of between 20x and 25x, a 6 inch (150 mm) telescope brings out a shade of the galaxy's signature dust lane, silhouetted, as it were, against the oval disk, which is oriented northwest to southeast. Some observers claim to be able to see it in a 4 inch (100 mm) telescope, but dark, clear skies are required. In the case of M 64, higher magnifications may be substituted for lower power since this will increase the contrast between the sky background, the brighter galaxy, and the dust lane.

Even without the distinctive dust lane, the Blackeye Galaxy would be considered unusual. Long-exposure photographs and digital images show tightly wound, softly textured arms that are nearly circular in shape. Even large ground-based telescopes are unable to resolve the smooth arms into knots of stars. In addition, a second, fainter dust lane can be seen in images on the inner edge of the luminous arm on the north side of the galaxy.

The Blackeye's rotation is also out of the ordinary. Recent observations appear to indicate that the galaxy's center rotates in the opposite direction to its outer disk. Astronomers suspect a collision between M 64 and two other galaxies may have set up the unusual internal motions.

Why the Blackeye?
The massive dark lane beneath the Blackeye's center is caused by an intervening cloud of dust silhouetted against the galaxy's bright nucleus.

Looking for a Blackeye
A small telescope picks up a suggestion of the Blackeye Galaxy's dust lane.

All Wrapped Up
The Kitt Peak National Observatory's 150 inch (4 m) reflector captured this image of the Blackeye. Note the galaxy's arms wrapped tightly around a smooth-looking core.

The Whirlpool Galaxy

- Constellation: Ursa Major
- Visibility: Northern Hemisphere and north of 36°S
- Magnitude: 8.1
- Apparent size: 11 x 8 arcminutes
- Distance: 23 million light-years

The lovely face-on spiral M 51, the Whirlpool Galaxy, lies just a little over 3 degrees south-southwest of Alkaid (Eta [η] Ursae Majoris), the magnitude 2 star marking the end of the Big Dipper's handle. In a 7 x 50 pair of binoculars, M 51 is a small blot of light, similar to a slightly out-of-focus star. Binoculars in the 10 x 50 and 20 x 80 range resolve this into a pearl-shaped object.

An 8 inch (200 mm) telescope at medium magnification reveals the galaxy's true nature—it is, in fact, a binary system. The main disk has a distinct bright round core set within a surrounding film of glaucous light that exhibits a hint of spiral structure. Just off the main disk to the north, we see a diffuse stellar point with flared regions of light on either side. This is NGC 5195, a more distant companion galaxy. Some 400 million years ago, these two galaxies passed near each other. The resulting gravitational tug-of-war pulled one of M 51's spiral arms out of round, as well as disturbing the structure of the companion.

Images made with the Hubble Space Telescope indicate that the galaxy's massive center is about 80 light-years across with a mass 40 million times that of the Sun. The density of stars in this region is about 5,000 times higher than in the Sun's neighborhood. The very core of M 51 contains about one million times the mass of the Sun confined to a region less than 5 light-years across. This, together with the pattern of crisscrossed dust lanes over the nucleus, suggest the presence of a massive black hole.

Two-in-One
The Whirlpool Galaxy (top) and its companion galaxy NGC 5195 are linked by a bridge of gas created by gravitational attraction. Hot, young stars color the spiral arms blue; stars are being born in the pinkish areas. This binary system is estimated to be more than 65,000 light-years across.

Ghost of Saturn
An early observer of the Whirlpool described the galaxy as "resembling the ghost of Saturn with its rings in a vertical position." Larger telescopes show a definite spiral structure.

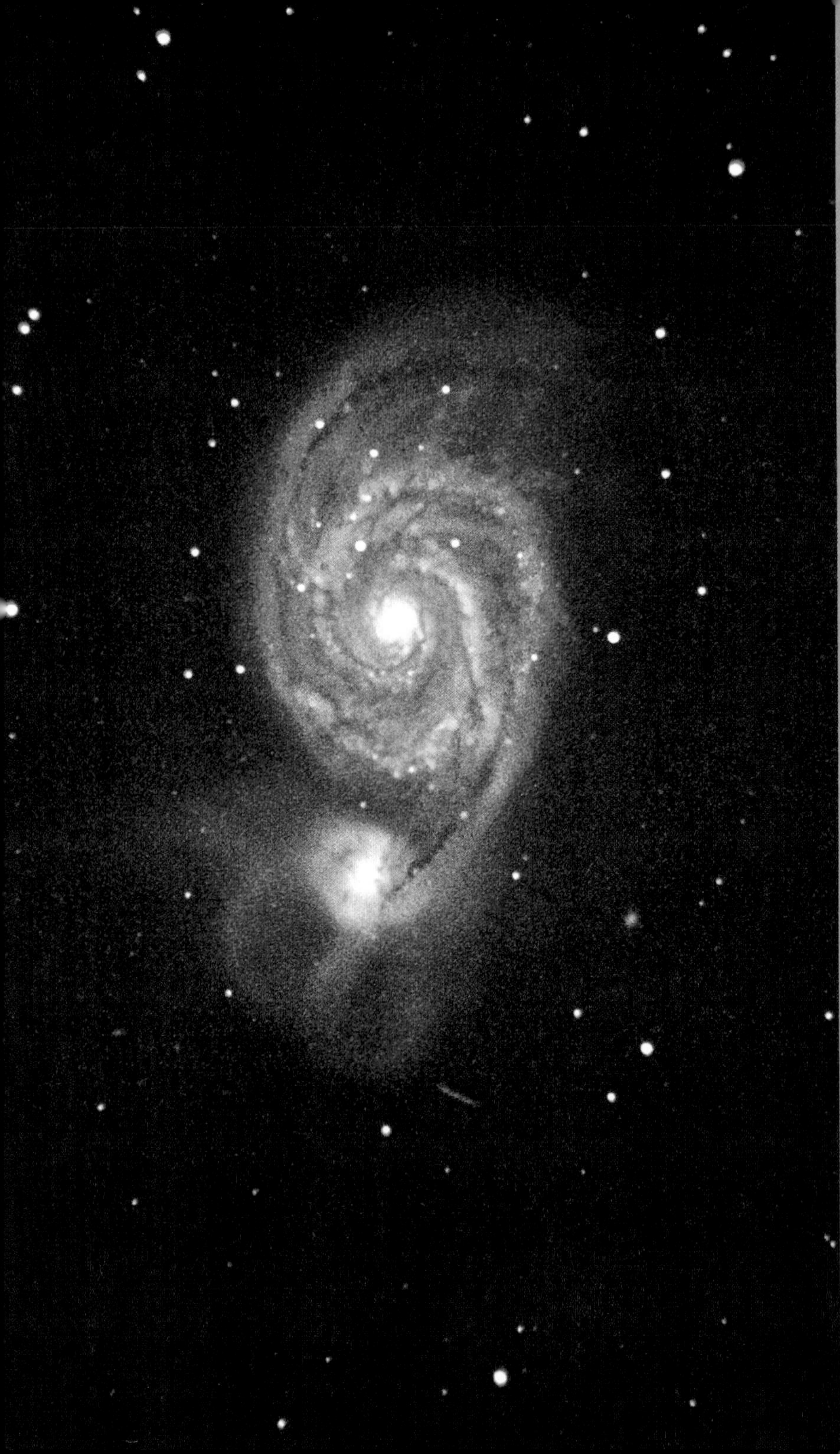

The Andromeda Galaxy

- Constellation: Andromeda
- Visibility: Northern Hemisphere and north of 42°S
- Magnitude: 3.4
- Apparent size: 3 x 1 degrees
- Distance: 2.4 million light-years

The Andromeda Galaxy (M 31) is far and away the brightest and largest galaxy visible from the Northern Hemisphere. At magnitude 3.4, it can be seen on dark evenings with the naked eye alone, some 3 degrees northwest of the middle of the Y-shaped constellation Andromeda. Here the eye catches upon a soft, "barely-there" glow, like a piece of detached Milky Way. Closer inspection with 7 x 50 binoculars reveals a decidedly oval disk of frosty light nearly 3 degrees in length—six times the apparent diameter of the Full Moon. At the center of the oval lies a brilliant starlike nucleus around which the light falls off smoothly toward the edge.

In dark skies, 11 x 80 or 20 x 80 binoculars show how really bright and huge M 31 is. If you use a telescope of any size, use the lowest power you have if you want to fit as much of it in the field of view as possible. In a 6 or 8 inch (150 or 200 mm) telescope, look carefully north and northwest of the galaxy's hub for hints of a curved dark lane.

Just south of the galaxy's bright center lies a separate oval patch of light visible as a 9th-magnitude fuzzy point in binoculars. This is the satellite galaxy M 32, an elliptical type and therefore smooth and featureless. On the opposite side of the disk, farther away, is the fainter galaxy M 110. This is another elliptical companion.

The Andromeda Galaxy is the nearest galaxy comparable to our own in size, and one of the few that happens to be approaching rather than receding. Still, even though it is considered nearby astronomically speaking, light from this galaxy takes 2.4 million years to reach Earth, making it one of the most distant objects visible to the unaided eye.

Naked-eye Sight
The Andromeda Galaxy is bright enough to be clearly visible to the unaided eye, given a dark enough sky. Here, it glows as a patch of fuzz near the bright stars of the constellation Andromeda. You are looking across 2.4 million light-years of space.

Galactic Companions
Andromeda is attended by two companion galaxies: M 110 (at left) and M 32, the starlike blob above Andromeda's center. The photo's field of view approximates that of binoculars or a telescope fitted with a low-power, wide-field eyepiece, but keep in mind that much more detail is shown here.

High-power View
Use a higher power eyepiece to zoom in on Andromeda's core.

409

The Pinwheel Galaxy

- Constellation: Triangulum
- Visibility: Northern Hemisphere and north of 53°S
- Magnitude: 5.7
- Apparent size: 67 x 41 arcminutes
- Distance: 2.9 million light-years

Fourteen degrees southeast of the Andromeda Galaxy lies another huge island universe, M 33, known as the Pinwheel. This face-on galaxy covers an area of 67 x 41 arcminutes, larger than the Full Moon's apparent diameter. Being so spread out on the sky, it has a correspondingly low surface brightness, and is therefore often difficult to locate.

M 33 is visible, just barely, in binoculars in a dark, clear sky. Nevertheless, many observers claim to be able to see this object with their unaided eyes, while others say they can barely discern it in the telescope. In any event, these conflicting reports should serve to alert novice observers that the Pinwheel Galaxy is probably going to be one of the more challenging objects they will pursue.

Once located, however, there are some fascinating visual aspects to this galaxy. In an 8 inch (200 mm) telescope at magnifications of 30x, the galaxy's nucleus is bright and globular—even less starlike than M 31 (see p. 408)—and surrounded by an attenuated patchy haze that gradually diminishes the farther you look from the center. In dark skies, 6 and 8 inch (150 and 200 mm) telescopes can pick out some hint of spiral structure in the contrasty sections in the disk, particularly on the north and west edges. Be sure to use averted vision (see How to See the Most, p. 111).

The keen observer's eye may also fall upon an enhancement in the disk 12 arcminutes northeast of the galaxy's hub. This is the star-forming region NGC 604. It is one of the largest stellar associations known, measuring over 1,000 light-years across. Its spectrum reveals that it is similar in composition to the Orion Nebula (see p. 394), but over 30 times larger.

Pink Spot
Apart from its white nucleus, the Pinwheel's most conspicuous feature is a pinkish spot of nebulosity, NGC 604, at outer left in this photo.

A Galactic Challenge
Although bright, at magnitude 5.7, the Pinwheel Galaxy is hard to see because it appears face-on and is spread over a large patch of sky.

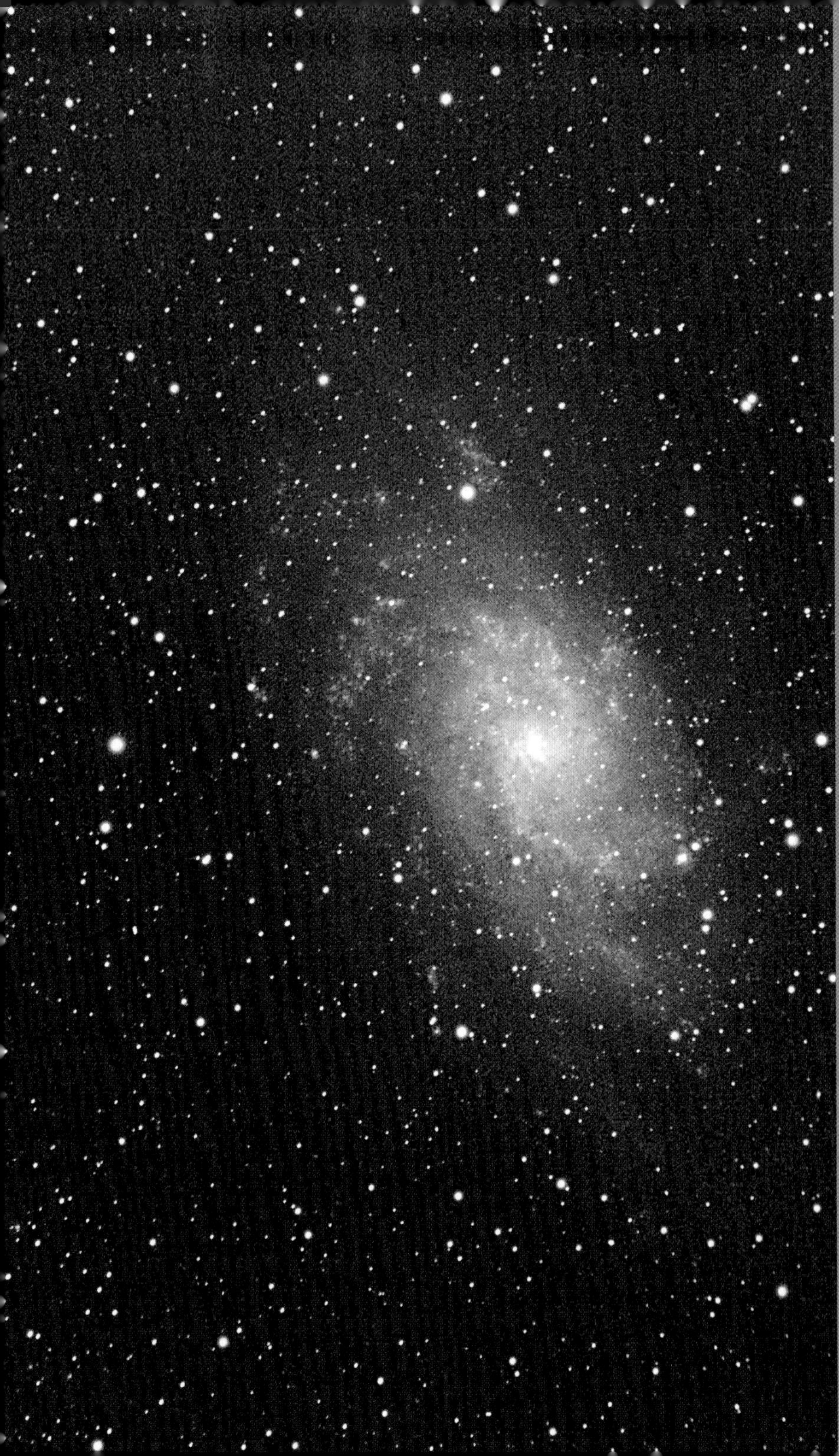

The Large Magellanic Cloud

- Constellation: Dorado
- Visibility: Southern Hemisphere and south of 10°N
- Magnitude: 1
- Apparent size: 7 x 7 degrees
- Distance: 163,000 light-years

Just 15 degrees and 36 degrees southwest of Canopus lie two bar-shaped patches of light that, to the naked eye, look for all appearances like broken-off pieces of the Milky Way. These clouds are not fragments of our Galaxy, however, but separate, nearby companion galaxies known as the Large and Small Magellanic Clouds (LMC and SMC).

Clouds of Magellan
The LMC is roughly twice the size of its companion, the SMC, and contains about 10 times its mass.

Because of their proximity to the south celestial pole, the Magellanic Clouds can be seen only from the Southern Hemisphere (though in late December and early January they may be glimpsed very near the southern horizon after sunset from latitude 10°N).

Both clouds belong to a class of galaxies called irregulars, because they exhibit no definite shape. The LMC is about 33,000 light-years in diameter and lies about 163,000 light-years away. It is rich in young, hot stars, star clusters, and bright nebulas. To view it, there is no need for large instruments, filters, high-power eyepieces, or special observing "tricks"—a pair of 7 x 50 binoculars or a small telescope at low magnification is all you need to enjoy this celestial treasure trove.

Tarantula in the LMC
The Tarantula Nebula (above) is the largest of the numerous nebulas in the Large Magellanic Cloud (facing page, below). These pinkish regions are sites of active star formation.

The most striking feature in the LMC is the Tarantula Nebula (NGC 2070), which is located at the southeastern end of the cloud. The nebula contains a group of about a dozen very hot, massive stars packed into a region less than 1 light-year across. This concentrated stellar bonfire lies at the center of spiderlike tendrils of hot gas that give the nebula its name. In 1987, a star exploded on the extreme western end of the Tarantula Nebula. Supernova 1987A was the first supernova seen with the naked eye since 1604 (see p. 364).

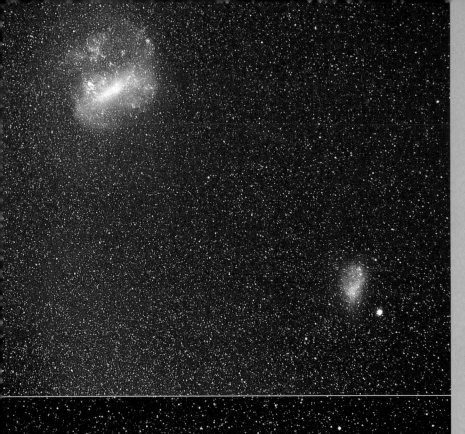

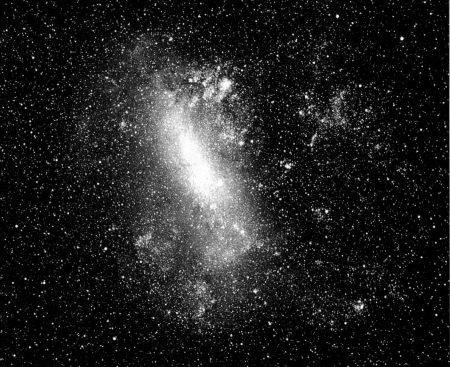

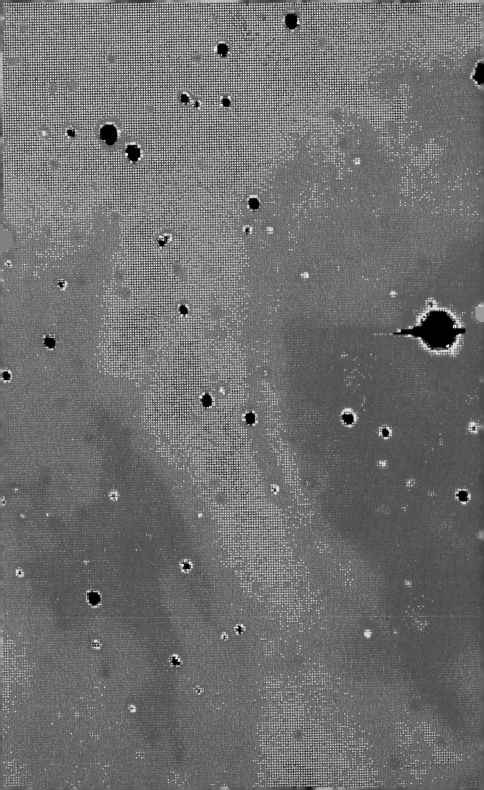

Contents

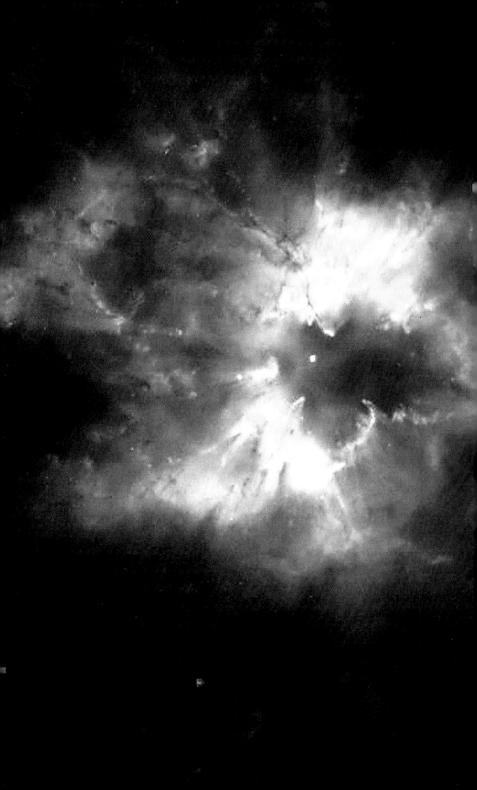

Introduction

Stars and planets pass overhead for all of us every night, but what we make of them is an individual matter. Some people are content to ignore the cosmos. Others, more sky-aware, want to explore this realm and make it a part of their lives.

This Guide is for all sky-aware people, but especially beginners. It has been designed to ease the way for anyone who wants to embark on a personal journey into the cosmos.

The book's first half is both an exploration of skywatching's fascinating past and an up-to-date guide to today's astronomy. In simple language it covers everything from the Earth and Moon to stars, nebulas, and galaxies. It also includes down-to-earth advice about choosing binoculars and telescopes, as well as helpful tips on viewing.

The second half of the book offers a field guide to the heavens. It gives you the what, where, and how of astronomical observation, so that you can discover the sky for yourself. The information is practical, easy to follow, and covers all kinds of observing, from the unaided eye to binoculars and telescopes.

An infinite universe lies within your reach every clear night. If you would like to get to know it better, you can start the adventure this evening.

How to Use This Guide

An armchair guide to astronomy past and present, and a practical observer's manual rolled into one, this Guide will help you journey among the planets, stars and galaxies.

Telescope Advice
The telescope entry on pp. 106–111 discusses the different kinds of telescopes on the market, how to purchase wisely, and what accessories you need to get the most out of owning a telescope.

The Guide is divided into two parts. Part One, *Discovering the Universe*, gives background on astronomy as a science, examines the main types of celestial bodies, and introduces the beginner to all aspects of backyard stargazing.

Part Two, *A Guide to Celestial Objects*, is very much practical in approach, and is meant to be taken outside and used in the field. It starts with our near-neighbor, the Moon, and works outward as far as the galaxies, the building blocks of the universe. Two chapters of star charts—one covering deep-sky objects, the

other, the monthly movements of the constellations—guide you through the night sky.

AN EXPANDING UNIVERSE

Hubble's All-seeing Eye in Space

With a mirror 94 inches (2.4 m) across, the Hubble Space Telescope (HST) is not the largest telescope in existence, but it has the best view of all. Orbiting 380 miles (600 km) above Earth, HST has observed all kinds of astronomical bodies, from the Moon to the most distant galaxies.

HST was launched in 1990 and is more than halfway through its planned lifetime. Scientists are now planning the Next

A shuttle astrona
protective c
Telescope's magne
regular m
upgrades from t
at

Part One pages provide an overview of astronomy, explaining its history and major concepts. Entries are arranged into three sections and seven chapters.

Frequent use of diagrams makes scientific information easy to understand.

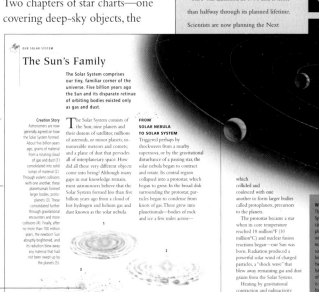

OUR SOLAR SYSTEM

The Sun's Family

The Solar System comprises our tiny, familiar corner of the universe. Five billion years ago the Sun and its disparate retinue of orbiting bodies existed only as gas and dust.

Creation Story
Astronomers are now generally agreed on how the Solar System formed. About five billion years ago, grains of material from a rotating cloud of gas and dust (1) consolidated into solid lumps of material (2). Through violent collisions with one another, these planetsimals formed larger bodies, proto-planets (3). These consolidated further through gravitational encounters and more collisions (4). Finally, after no more than 100 million years, the newborn Sun abruptly brightened, and its radiation blew away any material that had not been swept up by the planets (5).

The Solar System consists of the Sun; nine planets and their dozens of satellites; millions of asteroids, or minor planets; innumerable meteors and comets; and a plane of dust that pervades all of interplanetary space. How did all these very different objects come into being? Although many gaps in our knowledge remain, most astronomers believe that the Solar System formed less than five billion years ago from a cloud of hot hydrogen and helium gas and dust known as the solar nebula.

FROM SOLAR NEBULA TO SOLAR SYSTEM
Triggered perhaps by shockwaves from a nearby supernova, or by the gravitational disturbance of a passing star, the solar nebula began to contract and rotate. Its central region collapsed into a protostar, which began to grow. In the broad disk surrounding the protostar, particles began to condense from knots of gas. These grew into planetesimals—bodies of rock and ice a few miles across—

which collided and coalesced with one another to form larger bodies called protoplanets, precursors to the planets.

The protostar became a star when its core temperature reached 18 million°F (10 million°C) and nuclear fusion reactions began—our Sun was born. Radiation produced a powerful solar wind of charged particles, a "shock wave" that blew away remaining gas and dust grains from the Solar System.

Heating by gravitational contraction and radioactivity caused the protoplanets to melt and separate into the different internal layers we can observe in planets today. The first billion or so years was also a time of continuous bombardment by leftover debris, which battered the surfaces of the planets and their satellites. The craters of the Moon are mementos of this era.

WHEN WOR
The final sta
System's for
catastrophic
planet-size h
with its unu
may be the i
such inciden
between Ura
twice as ma
have caused
of Uranus' r
is now thou
formed from
when a bod
struck the y

THE NATURE
Because temp
with distance
composition
according to v
The inner sol.

120

Orion and Canis Major

Orion is star-forming territory. Long-exposure photographs and infrared images show that the entire constellation is wrapped in nebulosity—the stuff of new stars. The visible tip of the gas clouds is the Orion Nebula, a star factory some 1,500 light-years away. Trek east of Orion, along the main band of the Milky Way, and you enter star-cluster country.

Canis Major

Sky Tour 8

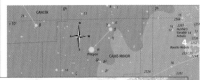

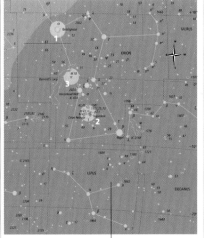

1 Betelgeuse (Alpha [α] Orionis) ◉ This red supergiant's name is a corruption of original Arabic words for "hand of al-jauza," an obscure term which may refer to a female character in ancient Arabic mythology.

2 M 42, the Orion Nebula (NGC 1976) ◉ ⊂⊐⊃ ⊂⊡⊃
Clearly visible to the unaided eye under a dark sky as a fuzzy star in Orion's Sword, M 42 blooms into a wondrous sight in any telescope. Great wings of gas curve away from a glowing mottled core, lit by a quadruplet star, the Trapezium star cluster.

Seen best in
Northern Hemisphere winter
Southern Hemisphere summer

Sky Tour 8

347

Graphic *feature pages* are scattered throughout Part One of the Guide to highlight topics of particular interest and visual appeal.

Part Two pages focus on the practical side of astronomy. There are two chapters of star maps, and five that describe how to observe specific celestial bodies.

The *star maps* are designed to be as straightforward as possible, allowing beginners to quickly develop their sky-watching skills.

Motion
nets of the
orbit the Sun
direction and
e same plane.
ception, their
its are very
ar. Pluto has a
cal orbit that
akes it inside
orbit.

Throughout Part One, *fact boxes* provide additional informa-tion—significant dates, practical tips, a quirky story—about the topic discussed on the page.

121

A Timeline of Astronomy

William Herschel's reflector

Pre-16th Century

32,000 BC ■ Lunar phases scratched on bone—oldest astronomical record?

4,000 BC ■ First records of the oldest constellations by Sumerians of Mesopotamia.

500 BC ■ Pythagoras says the heavens are made of crystal spheres; Earth is at the center.

350 BC ■ Aristotle describes the phases of the Moon and how eclipses happen.

300 BC ■ Aristarchos proposes a Sun-centered model for the universe.

200 BC ■ Eratosthenes accurately measures the circumference of Earth.

150 BC ■ Hipparchos creates magnitudes to measure star brightness, discovers precession, and compiles the first star catalog.

AD 150 ■ Claudius Ptolemy publishes a summing-up of the ancient world's astronomical knowledge.

16th & 17th Centuries

1543 ■ Nicholas Copernicus proposes a Solar System centered on the Sun.

1576 ■ Tycho Brahe begins compiling accurate observations of the motions of stars and planets.

1608 ■ The telescope is invented by Hans Lippershey.

1609 ■ Johannes Kepler uses Tycho's observations to find the orbit of Mars is elliptical.

1610 ■ Galileo Galilei discovers moons orbiting Jupiter, craters on the Moon, and stars in the Milky Way.

1619 ■ Kepler discovers a mathematical relationship between the length of a planet's year and its distance from the Sun.

1687 ■ Isaac Newton publishes the *Principia*.

18th & 19th Centuries

1759 ■ Comet Halley makes the first predicted return of a comet.

1781 ■ William Herschel discovers Uranus.

1846 ■ Johann Galle and Heinrich d'Arrest find Neptune, using predictions by Urbain Leverrier and (independently) John C. Adams.

1860s ■ The spectroscope shows astronomers what stars and nebulas are made of.

1880s ■ Photography becomes an important tool in astronomy.

Isaac Newton's telescope

20th Century

1905 ■ Albert Einstein publishes his special theory of relativity.

1908 ■ Ejnar Hertzsprung divides star populations into giants and dwarfs.

1912 ■ Henrietta Leavitt discovers that Cepheid variable stars with short periods of variation are dimmer than those with longer periods.

Vesto Slipher finds "spiral nebulas" receding from Earth, the first detection of the expanding universe.

1916 ■ Einstein publishes his general theory of relativity.

1917 ■ 100 inch (2.5 m) Hooker telescope erected at Mount Wilson, California.

1923 ■ Edwin Hubble shows that "spiral nebulas" are galaxies lying beyond the Milky Way.

1929 ■ Hubble's observations show the universe is expanding and provide estimates of its age and expansion rate.

1930 ■ Clyde Tombaugh discovers Pluto.

1932 ■ Karl Jansky discovers radio waves from space.

1937 ■ Grote Reber finds radio waves coming from the center of the Milky Way.

1938 ■ Hans Bethe explains how nuclear reactions cause the Sun to shine.

1946 ■ John Hey and others identify the most powerful radio source in the sky, Cygnus A.

1948 ■ 200 inch (5 m) Hale telescope is completed on Palomar Mountain, California.

George Gamow, Ralph Alpher, and Robert Herman describe how the chemical elements were formed in the Big Bang.

1952 ■ Walter Baade announces that galaxies are twice as far away as astronomers had assumed.

1957 ■ Launch of Sputnik 1 by the Soviet Union starts the Space Race.

1959 ■ First photographs of the Moon's farside.

1961 ■ The Soviet Union's Yuri Gagarin is the first man in space.

1962 ■ Mariner 2 discovers Venus' heavy atmosphere and hot surface.

1963 ■ Maarten Schmidt discovers that quasars lie very far away.

1965 ■ Arno Penzias and Robert Wilson discover microwave radiation left over from the Big Bang.

Mariner 4 is the first spacecraft to fly past Mars.

1967 ■ Discovery of pulsars by Jocelyn Bell-Burnell. These are soon identified as neutron stars.

1969 ■ Neil Armstrong and Edwin Aldrin make the first manned landing on the Moon (Apollo 11).

1973 ■ First flyby of Jupiter, by Pioneer 10.

1974 ■ First close-up photos of Venus' cloud tops and Mercury's heavily cratered surface, by Mariner 10.

1975 ■ First photos from the surface of Venus, by Venera 9.

1976 ■ Viking 1 and 2 land on Mars in an unsuccessful attempt to detect life.

Sojourner rover

1977 ■ Discovery of the rings of Uranus.

1978 ■ James Christy discovers Chiron, moon of Pluto.

1979 ■ Voyager 1 and 2 fly past Jupiter, discovering its rings.

First flyby of Saturn, by Pioneer 11.

1980 ■ Alan Guth proposes the early universe expanded extremely fast in a process called cosmic inflation.

Magellan

First detailed study of Saturn, by Voyager 1.

Very Large Array radio telescope starts operations in New Mexico.

1983 ■ Infrared Astronomical Satellite (IRAS) surveys the infrared sky.

1986 ■ First flyby of Uranus, by Voyager 2.

1987 ■ Supernova 1987A appears in the Large Magellanic Cloud.

1989 ■ First flyby of Neptune, by Voyager 2.

Margaret Geller and John Huchra announce that galaxies clump into "walls" and "voids" in the universe.

1990 ■ Hubble Space Telescope launched.

Magellan spacecraft begins radar mapping of Venus.

1991 ■ Galileo spacecraft on the way to Jupiter makes first asteroid flyby, 951 Gaspra.

Launch of Compton Gamma-Ray Observatory.

1992 ■ COBE satellite discovers "lumps" in the cosmic background radiation.

First Keck 400 inch (10 m) telescope commissioned on Mauna Kea, Hawaii.

1994 ■ Comet Shoemaker-Levy 9 crashes into Jupiter.

1995 ■ Galileo arrives at Jupiter; sends probe into atmosphere and begins tour of planet and moons.

1996 ■ Milky Way found to have a massive black hole at its center.

1997 ■ Mars Pathfinder spacecraft lands on Mars with Sojourner rover.

1998 ■ Expansion of the universe found to be accelerating.

1999 ■ Launch of Chandra, X-ray satellite observatory.

2000 ■ Discovery of water seeps and widespread sedimentary deposits on Mars.

Iapetus

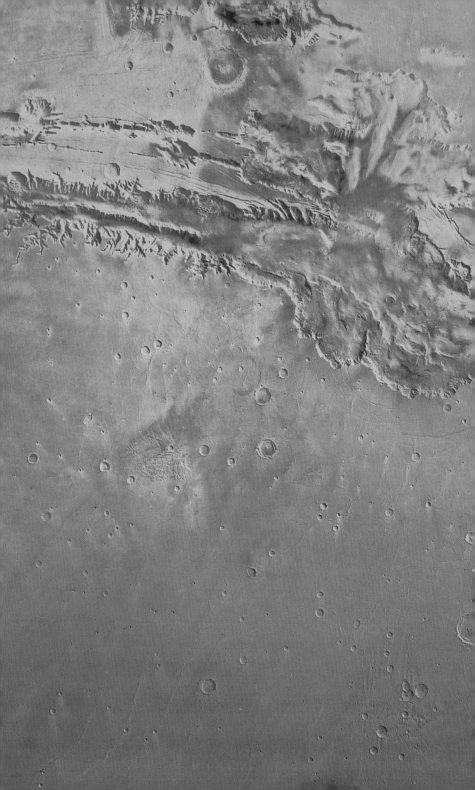

PART ONE

Discovering
the Universe

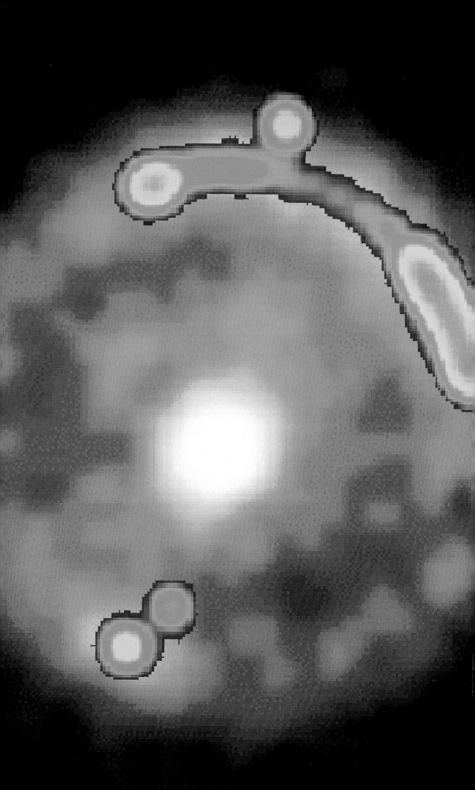

Searching the Heavens

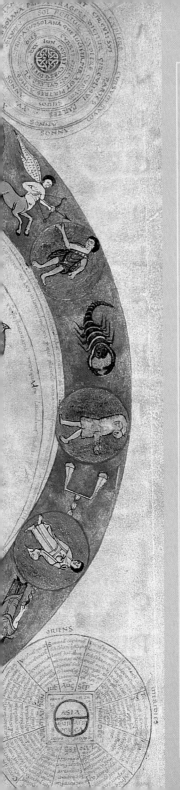

Earth in
the Cosmos

For many thousands of years, people believed that Earth lay at the center of the cosmos and that the heavens were the domain of the gods. This chapter charts human-kind's slow and painful awakening to a more rational understanding of its place in the universe.

"In the beginning, the gods Anou, Enlil, and Ea divided everything up between the two gods who were the keepers of the sky and earth... Sin and Shamash were given two equal portions, day and night."

From *Enuma Elish*, Assyro-Babylonian tablets 7th century BC

What Is Astronomy?

Astronomy, the science of studying the sky, has come far from its beginnings tens of thousands of years ago. Yet the urge to understand the universe runs as strongly in us as it did in our remote ancestors, even as the tools and techniques used in that quest have changed out of all recognition.

I n earliest times, astronomy (from the Greek meaning "the classifying of stars") was indistinguishable from astrology ("knowledge of the stars"): people looked to the sky for signs from the gods, while astronomer–priests consulted the heavens for portents. Today the two pursuits could not be farther apart, although ironically "knowledge of the stars" is actually an excellent description of modern astronomy.

Starting Out
Astronomy is an activity for a lifetime of learning and wonder. Starting is as simple as watching a beautiful grouping of the Moon and planets or noting the progress of constellations through the seasons. Binoculars open up new vistas, from the Milky Way resolved into myriad stars to the shuttling moons of Jupiter.

Images from Space
Images such as this, of filaments of gas in the Tarantula Nebula, 160,000 light-years away, are humbling reminders of the immensity and beauty of the universe.

BIG QUESTIONS
Nothing characterizes astronomy better than its relentless focus on big questions and issues. How did the universe form? What is its ultimate fate? What makes stars, galaxies, and planets? Are there other Earths out there? And, perhaps the biggest question of them all, Are we the only intelligent life in the universe?

Even if we get close to answering some of these questions—and our current level of knowledge is extraordinary compared to that of previous generations—our understanding of the universe will always be imperfect. The universe is simply too big and its workings too complex for us to come to any definitive conclusions. But, for astronomers, it is exactly that open-ended approach to

important issues which makes this science such a challenging and fascinating subject.

TWO ASTRONOMIES
Today's astronomy exists in two realms. One belongs to the professional astronomer. Steeped in physics and mathematics, astronomers use spacecraft, telescopes, and computers to probe the unknown corners of

the universe. Complex theories come and go as they are tested against observations and found wanting. A catalog of astronomers' discoveries in the 20th century alone would fill volumes; they range from a new planet (Pluto) in our Solar System to new solar systems far from us.

A second realm of astronomy lies as close as your own backyard. Thanks to the work of their professional counterparts, amateur astronomers have at their fingertips an understanding of the universe and its workings far beyond anything even dreamt of by ancient skywatchers. As well, recent years have seen a tremendous growth in telescopes and other equipment specially built for use in the backyard. Truly, there has never been a better time to take up skywatching.

THE NEW COSMOS

Astronomy stands today on the brink of great discoveries. Space probes, increasingly sensitive instruments, and new techniques will help humanity push back the bounds of the known universe and reveal much about how the cosmos works. If the 20th century is anything to go by, the 21st will be a time of profound and, perhaps, revolutionary change in astronomical practice and knowledge.

Travelers in Space
Probes to the planets, such as Cassini, now on its way to a rendezvous with Saturn, have greatly enhanced our understanding of the Solar System.

The First Astronomers

Astronomy's roots lie in the stargazing done by our Ice Age ancestors some 30,000 years ago. Living by hunting and gathering, early humans tracked stars as well as game, and foretold changes in the seasons by changes in the sky.

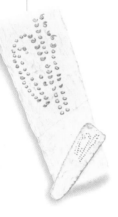

The First Watchers
Much archaeological evidence suggests that humans have long been fascinated by the heavens. This 32,000-year-old shard of bone (above), found in France, is marked with what may be a track of the Moon's phases. Stonehenge (below) was begun more than 4,000 years ago as a monument to mark sunrise on the solstice.

We can only guess at when our ancestors began trying to understand the sky. In Europe, archaeologists have found what may be lunar calendars that were carved onto bone more than 30,000 years ago during the last Ice Age. Dating from more recent prehistoric times are numerous celestially aligned monuments in both the Old World and the New, which leave no doubt that ancient people tracked the Sun, Moon, and stars.

Living close to nature, people noted the daily path of the Sun and its year-long traverse of the seasons. The Moon's quicker movement marked shorter rhythms in the year, as it once did for North American Indians:

they saw the Full Moon during our February as the "Hunger Moon," while the one that fell in July was the "Thunder Moon" or "Hay Moon." The motions of the Sun and Moon, as well as the stars, provided a clock and calendar, rough by our standards, but adequate for the needs of these hunter-gatherers.

HOME OF THE GODS

The sky had other uses as well. Although archaeological evidence is scanty, it seems likely that to early peoples the heavens were the domain of the gods and that the movements of the heavenly bodies had religious explanations. Humans see patterns in the world around them, and it was natural

to organize the sky's scattering of random stars into meaningful groups reflecting those beliefs.

Also, at a time when no writing existed, people had to commit all knowledge to memory. A storybook in the sky helped everyone recall the legends that gave meaning to life. The earliest constellations—arbitrary groupings of stars—surely grew from such mythologies, but the identities of the first celestial figures that paraded over the world have vanished with the peoples who first named them.

CONSTELLATIONS TAKE SHAPE

Skywatching took a leap forward when settled agricultural societies arose about 10,000 years ago in Mesopotamia, the fertile land between the Tigris and Euphrates rivers in what is now Iraq. Since farming is governed by the seasons, knowledge of the sky's unceasing rhythms became of even greater importance as a way of determining the best times to plant and reap.

The oldest surviving constellations probably date from these times. Figures we now

know as Leo, Taurus, and Scorpius begin to appear in Mesopotamian carvings from the 3rd millennium BC. These constellations marked significant places in the Sun's apparent path in the heavens— where it rose and set due east and west, and the farthest limits of its journeys north in summer and south in winter. Such locations were important turning points in the farming year.

Early skywatchers also noticed that the Sun and Moon appeared

Circles in Stone

Carved into stone at a Native American site in New Mexico are concentric circles like this one. One theory is that they may be stylized impressions of a supernova that took place around AD 1200.

Babylonian Tablets

Astronomers in ancient Babylonia painstakingly recorded the movements of stars and planets on clay tablets more than 2,500 years ago. Such records were mines of information for later civilizations, including the Greeks.

BIG HORN MEDICINE WHEEL

Among the world's lesser-known solar monuments is Wyoming's Big Horn Medicine Wheel. While its exact function, builders, and construction date are unknown, the spokes of the wheel are oriented toward the summer-solstice sunrise and sunset. They may also be sightlines to the points on the horizon where stars rise.

The Sun as Deity

As the most obvious and impressive of the heavenly bodies, the Sun was greatly revered by early civilizations. In most cases, the chief deity was the Sun god. The carving on this 3,000-year-old bronze vessel from ancient Persia depicts the Sun god at a banquet.

to travel among 12 prominent constellations, later collectively called the zodiac (see p. 28). This, they decided, was the home of the Sun and Moon deities. In addition, five special "stars," each believed to be the home of a god, also traveled the zodiac. Today these are known to be planets, a word that derives from the Greek for "wanderer."

The zodiac was also where eclipses occurred—rare, greatly feared events in which the Moon turned an ominous copper color and the Sun's light was shut off for what must have seemed an eternity to those watching.

SEEING OMENS IN THE SKY

Archaeological discoveries show that it was also in Mesopotamia that astronomer–astrologers first appeared. This priestly caste was dedicated to the study of the night skies to seek omens for their ruler.

Sumeria was the first great Mesopotamian civilization, arising in the 4th millennium BC. The Sumerians devised the plow, wheeled vehicles, major irrigation projects, and writing. They also amassed a significant amount of skylore, which passed to their successors, Babylonia and Assyria.

Using the Sumerian legacy, the Babylonians and Assyrians developed a sophisticated understanding of the sky and its patterns. They created calendars for planting crops and had the ability to accurately predict eclipses of the Moon. The Babylonians invented the degree for measuring angles.

From Mesopotamia, much astronomical knowledge passed to the Greeks relatively unchanged (see p. 30). They adopted the degree, for example, and imported such familiar constellations as Auriga, Gemini, Leo, Capricornus, and Sagittarius. Their Mesopotamian names were simply translated into Greek.

RITUAL STARS

Ancient civilizations outside
Mesopotamia also developed
their own skylore. In Egypt, the
regular floods of the Nile River
controlled all life by irrigating
and fertilizing the fields. Egyptian
astronomer–priests used the rising
of the star Sirius before the Sun
to predict these floods. The sky
reflected Egyptian gods—our

The Egyptian Cosmos

Early societies often saw
the sky as an immense
vault. This engraving from
an Egyptian mummy case
shows Shu, the god of the
atmosphere, lifting up his
daughter Nut, goddess of
the sky, to separate her
from Earth.

Gifts to the Sun

Religion and ritual linked
humans—and royalty in
particular—to the heavens.
In this stone relief, the
Egyptian pharaoh
Akhenaten (1352–36 BC)
presents lotus flowers to
the Sun disk Aten. Each
pharaoh was customarily
referred to as "son of the
Sun": he ruled the social
order while the Sun ruled
the cosmic order.

25

and astronomy: the emperor was seen as forming a personal link between heaven and Earth.

Astronomers sought to tell the future and, by rituals, to minimize bad omens and enhance good ones.

THE NEW WORLD

In the New World, astronomer–priests of the Maya and Aztec civilizations made extensive observations of heavenly bodies. The Maya inhabited southern Mexico, and flourished between the 3rd century BC and the 9th century AD. They created a cosmology from recurring patterns among the stars and planets, particularly Venus, which they associated with the god of rain.

To the Aztecs, who dominated what is now central Mexico for two centuries before being conquered in AD 1520 by the Spanish, Venus represented the god Quetzalcóatl. This feathered serpent emodied the power of life emerging from earth, water, and sky. Rituals and blood sacrifices

Chinese Astronomy
Chinese astronomers set down the rhythms of the heavens in documents such as this Tang Dynasty disk. From the center outwards, it shows the solstices and equinoxes, the zodiacal cycle, and the "mansions" through which the Moon passes each month.

Orion was Osiris and the Milky Way represented the goddess Nut giving birth to the Sun god Ra.

Chinese astronomers accurately observed stars, planets, supernovas, and comets. In about 1300 BC, they produced what may be the world's first calendar. As in Egypt and Mesopotamia, there was no distinction between astrology

SAME STARS, DIFFERENT FIGURES

Star patterns from Mediterranean mythology, supplemented by modern inventions, have become the official constellations of the sky. But different cultures see different patterns in the same stars. Ancient Egyptians recognized Orion as the god of light, while today some Amazonian Indians see the same stars as a giant crocodile. Among the Inuit of Canada's Arctic, the stars of the Big Dipper are seen as caribou. In sea-going Greenland, however, these stars become a kayak

stand. To the Aztecs, the Dipper was the god Tezcatlipoca, who stirred up trouble as he circled the pole. But bright stars are not always important. The Aborigines of northern Australia saw a crab in five faint stars in Hydra, ignoring bright Procyon and Regulus nearby. The Quechua Indians of Peru, direct descendants of the Incas, recognize dark constellations made from patches of interstellar dust in the Milky Way. They see the Coalsack Nebula as a dove, while a baby llama lies in a dark cloud near Alpha Centauri.

were required to appease it as Venus disappeared and reappeared over a cycle of five apparitions in eight years.

To the island cultures of the Pacific Ocean, astronomy was very much a practical art. The Gilbert Islands of Micronesia, for example, have no word for astronomer. Instead, skylore lay in the keep of navigators. In the tropics, stars and planets rise and set at steep angles to the horizon. Thus a bright object makes a beacon to steer by for hours without risk of going astray. Navigators learned sky stories that gave sailing directions, and used stick-and-shell maps showing islands and wave patterns.

Stone Serpent
Quetzalcóatl, the feathered serpent, was the most important of the Aztec gods. In his various forms he embodied the Sun, the wind, and Venus.

Aztec Calendar
This "stone of the Sun" depicts the Aztec calendar. At its center is the Sun god Tonatiuh, who demanded offerings of human blood.

Stellar Navigators
For centuries, the people of the Marshall Islands in the Pacific used simple devices for navigation embodying hard-won skylore.

The Zodiac:
Signs in the Sky

Earth's yearly orbit makes the Sun appear to glide through 12 constellations known collectively as the zodiac, from the Greek word meaning "life." The zodiac's first constellation is Aries the Ram, where the Sun crosses into the northern celestial hemisphere at the start of northern spring.

The zodiacal signs acts as markers in astrology—the belief that the stars and planets influence human character traits and daily activities. Early astronomer–astrologers focused particularly on the interplay between the positions of planets and the zodiac's constellations. Although astrology continues to fascinate some people, there is not one shred of scientific evidence that the way constellations and planets line up has ever affected anyone.

Chinese Zodiac
The 12 creatures of the Chinese zodiac are the ox, rat, dog, tiger, dragon, monkey, horse, sheep, rooster, snake, rabbit, and pig.

Turkish Twins
This depiction of Gemini comes from a 16th-century Turkish literary text.

Circle of Signs

The signs of the zodiac dominate this 16th-century Italian engraving of the universe, which has Earth at the center.

Years, Animals, and Personality

Each Chinese zodiacal animal, three of which are represented here as figurines, embodies a personality type. People born into a year governed by the ox (lower right) are believed to be likable, hard working and thrifty; those born in years of the sheep (upper right) are kindly and emotional; and people born in years of the dog (far right) are intelligent and value fairness and justice.

Greek Astronomy

The ancient Greeks made the first great advances in astronomy. Observing the heavens, a succession of brilliant thinkers used geometrical principles rather than supernatural beliefs to explain what they saw. In so doing, they broke the link between astronomy and astrology.

Celestial Spheres
The universe consisted of a series of crystalline spheres, according to the philosopher–mathematician Pythagoras. In this detail from Raphael's fresco, *The School of Athens*, he is seen at left engrossed in calculations while other philosophers look on.

Like most early agricultural people, the earliest Greeks observed the sky and used its movements to set the pace of their yearly round of farming activities. And like other people, they created and named constellations—the earliest perhaps between 3000 BC and 2000 BC—filling the sky with a storybook of mythology that gave everyone a nightly reminder of the gods and heroes.

At this early period, much skylore was probably borrowed from the civilizations of Mesopotamia (see pp. 23–4). And like its Mesopotamian forerunners, astronomical

knowledge among the prehistoric Greeks was likely weighed down with religious symbolism and astrological omens.

THE GREEK REVOLUTION BEGINS
At first, Greek astronomy concerned itself with purely practical purposes. The poets Homer and Hesiod wrote about astronomy in such terms during the 8th century BC. The Homeric heroes Achilles and Odysseus certainly regarded the Pleiades, Orion, Taurus, Boötes, the Great Bear, and the star Sirius as familiar objects to help with navigation and timekeeping. And Hesiod's rustic poem *Works and Days* describes an agricultural calendar controlled by the rise and set of various constellations and stars.

But from the 6th century BC onward, Greek thinkers broke with both practical astronomical concerns and the mythologies of the past. They went beyond metaphysical explanations for the movements of the heavens to propose reasons based on geometry and mathematics. In the 800 years until the death of Ptolemy in AD 150, the basis of modern astronomy came into being. The leaders of this revolution were Greeks living in Ionia (Asia Minor) and southern Italy, whose ideas were enriched

THALES
Septem sapientum primus collaudatus
Apud Achillem Maffium

by contacts through trade with the astronomy and mathematics of Mesopotamia and Egypt.

Thales (6th century BC) went to Egypt to study mathematics. He is said to have predicted a solar eclipse and claimed Earth to be spherical.

Another Ionian was Pythagoras (6th century BC), a geometrician and mystic who proposed that the universe was made of concentric crystalline spheres surrounding Earth, nested like Russian dolls. The Sun, Moon, planets, and stars each traveled on its own sphere. Pythagoras also believed that these produced a haunting "music of the spheres"

as they rubbed against each other. In the 4th century BC, Eudoxos adopted Pythagoras' spheres and added more. These helped account for the irregular movements of the planets and the Moon which were apparent even with the crude measurements of the time.

MORE BREAKTHROUGHS— AND A RADICAL PROPOSAL

The philosopher Aristotle (4th century BC) wrote on many subjects and was greatly influential. He showed that Earth was a sphere, but he remained convinced that it was at the center of the universe because he could not see the stars shift position during the year, which they should if Earth moved around the Sun. (In fact the stars *do* shift, but the amount is too small to have been detected by the instruments of the day.)

Thales' Prediction
According to the Roman historian Herodotus, the Greek astronomer Thales predicted an eclipse of the Sun in 585 BC.

Aristotle's Legacy
Aristotle produced learned works on such diverse subjects as physics, botany, politics, ethics, and art, as well as astronomy. His Earth-centered ideas about the nature of the universe were to hold sway for 2,000 years.

CENTURIES OF GREEK THINKERS
6th century BC	Thales and Pythagoras
4th century BC	Eudoxos and Aristotle
3rd century BC	Aristarchos
2nd century BC	Hipparchos
2nd century AD	Ptolemy

Lasting Achievement
Among his many achievements, Hipparchos invented a system of magnitudes for expressing the brightness of stars. In modified form, it is still in use today.

Ptolemy
Centuries of Greek scientific tradition culminated in the work of Ptolemy, the last of antiquity's great astronomers. Though complex—and Earth-centered—his system accurately accounted for the movements of the heavenly bodies.

In the 3rd century BC, Aristarchos proposed a break with tradition—a universe with the Sun at its center—but his radical cosmic model was scorned and then forgotten. He also tried to measure the size of the Sun and Moon and their distances from Earth, but his effort failed because contemporary instruments were too coarse.

The greatest astronomer of ancient times was Hipparchos (2nd century BC). He accurately measured the distance from Earth to the Moon, reaching a figure of 29.5 Earth diameters, very close to the modern value of 30. He also compiled the first known star catalog, and devised the still-used magnitude system for comparing star brightnesses. But Hipparchos' greatest finding came when he examined older observations of stars. Comparing what he could see with old records, he noticed differences and out of those changes he discovered precession, a slow wobble in Earth's axis caused by the gravitational pull of the Sun and Moon.

PTOLEMY'S CYCLES AND EPICYCLES

The last great Greek astronomer of antiquity was Claudios Ptolemaios, known as Ptolemy, who lived in Alexandria, Egypt, in the 2nd century AD. Building on Hipparchos' observations, Ptolemy developed a detailed mathematical theory to predict the movements of the Sun, Moon, planets, and stars—all using the Earth-centered model of the universe. His work was published in a book now known by its Arabic title of *Almagest*.

As Ptolemy and most of antiquity's astronomers saw it, the perfect bodies in the heavens had to follow the only perfect shape, a circle. The Ptolemaic system therefore demands that all celestial objects move at constant speed in circular orbits. Even coarse observations show this is false,

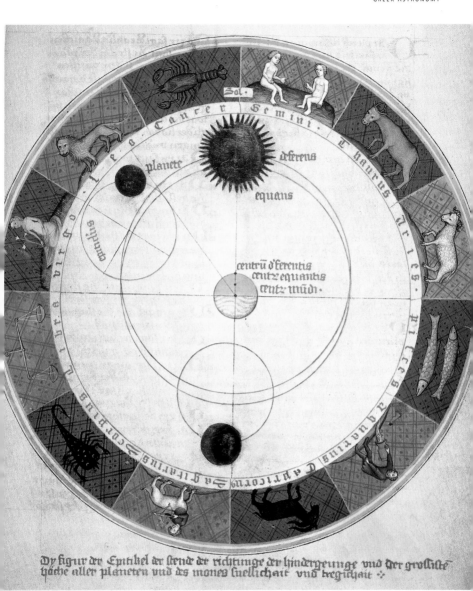

Ptolemy's Universe
A 16th-century engraving shows Ptolemy's scheme: each planet moves around Earth in a small circle, or epicycle, on the rim of the main circular orbit.

so Ptolemy gave his orbits epicycles—smaller circular orbits on top of the main orbit—plus he freed some orbits from being centered on Earth.

His scheme certainly worked, although to us, Ptolemy's universe looks contrived. Even a non-astronomer can sense that it rejects physical reality for the sake of a mathematical ideal. Despite this, Ptolemy's cosmic model made remarkably accurate predictions, and in the end that was what antiquity's astronomers were chiefly seeking.

The Telescopes that Opened Up the Universe

In late 1609, word reached the Italian scientist Galileo Galilei of a Dutch invention—two lenses in a tube—that let an observer see distant things up close. Grinding his own lenses, he made a small telescope which he turned on the sky. This refractor was cruder than even a cheap pair of binoculars today, but it opened a new world as Galileo made discovery after discovery.

A new type of telescope, a reflector, was developed by Isaac Newton in England in 1672. With the lens replaced by a mirror, several bothersome optical defects in refractor telescopes were removed, paving the way for the giant reflector telescopes of today.

The Moon in Focus
Earlier observers had believed the Moon's surface to be perfectly smooth. Now Galileo showed that the Moon was in fact scattered with mountains and craters. These sketches were published in his ground-breaking book, *The Starry Messenger*, in 1610

Galileo's Telescopes
The telescope Galileo used to observe the Moon had a convex lens at the front of the tube and a concave one at the other end for an eyepiece. Two of his telescopes are shown here, together with the front lens from another.

Hawaiian Giant

Most modern telescopes are reflectors—
direct descendants of Newton's first primitive
model. Shown above is the protective dome
that houses one of two 400 inch (10 m) Keck
reflectors at Mauna Kea crater, Hawaii.

The Power of Mirrors

Isaac Newton's telescope was very small,
with a mirror just over an inch (25 mm) in
diameter and a focal length of just 6 inches
(150 mm). Yet it reportedly showed details
similar to those revealed by a refractor 3 to
4 feet (about a metre) long and with a
magnifying power of 40 times.

Copernican Astronomy

After the death of Ptolemy around AD 150, Western astronomy fell asleep for more than a thousand years. Aside from the work of Arab astronomers, virtually nothing new happened in astronomy until the 1500s when one European astronomer proposed a radical theory that broke with the past.

Revolutionary Views
Copernicus' Sun-centered system changed humanity's view of the cosmos.

The Great Observer
Tycho Brahe left a legacy of accurate observations that his assistant, Johannes Kepler, would exploit.

An Arabic empire arose in the middle of the 7th century AD, and astronomy flourished for centuries within it. Ptolemy's work was translated and studied by Arab astronomers, who refined it using accurate observations made with improved instruments.

Although the Arabs broke no new ground, they were for many years far in advance of sky-watchers in the Christian world. There, the Church had adopted Ptolemy's model and turned its attention to protecting and expanding its control over theology, politics, and learning.

The man who set astronomy moving again was Nicholas Copernicus (1473–1543). Born in Poland, Copernicus joined the Church after studying astronomy and mathematics. Driven perhaps by dissatisfaction with the artificial nature of the Ptolemaic system, he decided to revise Ptolemy. His grand synthesis, *On the Revolutions of the Celestial Spheres*, was published in 1543.

In his system, Copernicus kept many Ptolemaic features such as circular orbits and epicycles, but he made a revolutionary change by placing the Sun at the center of the universe and making Earth and the other planets orbit it. Surrounding all lay a vast sea of stars.

TYCHO BRAHE

Copernican ideas spread slowly. Converts were few at first, and an important doubter was the Danish astronomer Tycho Brahe (1546–1601). As a young man, Tycho saw that the ancients' ideas of an unchanging cosmos were wrong, and he set out to make the most accurate observations ever made with the idea of re-establishing astronomy from scratch. On a Baltic island, he built Uraniborg ("Star Castle"), Europe's most advanced observatory, where from 1576 to 1597

The following labels appear on the engraving:

PLANISPHÆRIVM
Sive
VNIVERSI TO
EX HYPO
COPERNI
PLANO

COPERNICANVM
Systema
TIVS CREATI
THESI
CANA IN
EXHIBITVM

he amassed a wealth of precise observations far more complete than anything antiquity had

accomplished. He later settled in Prague, and developed a cosmological model that blended Ptolemy and Copernicus. In Tycho's universe, the planets orbit the Sun while the Sun orbits a stationary Earth.

SEEKING HARMONIES

Yet despite his accurate observations, Tycho's hybrid model of the heavens attracted fewer converts than Copernicus' model had. In Prague, Tycho hired an assistant, the Austrian astronomer–mathematician Johannes Kepler (1571–1630). Kepler was already a Copernican, part of a growing number, and he possessed a fertile mathematical imagination that let

New and Old
A 17th-century engraving shows the Copernican system, with the Sun orbited by the planets. Modern though this may seem, Copernicus still clung to the old view that orbits were perfect circles.

Mapping the Heavens
Working before the invention of the telescope, Tycho Brahe depended on instruments like this, an equatorial armillary, which allowed a celestial object's coordinates to be taken.

The Circle Is Banished
Johannes Kepler made a decisive break with the past by rejecting the idea of circular orbits and substituting the ellipse.

Heavenly Harmonies
Kepler believed that a divine geometrical harmony could explain the relative positions of the planets. A diagram from his book, *Cosmographic Mystery* (1596), shows the Solar System as a nest of spheres separated by the five solids of classical geometry.

him discern patterns in numbers. He believed that these could reveal the secrets of the universe because underlying all nature were harmonies that only mathematics could express.

Kepler struggled with Tycho's observations of Mars, trying to fit its motion into a Copernican circular orbit. Again and again he failed. Kepler's great leap of imagination was to abandon the circular orbit and substitute an ellipse, which fit the observations perfectly. The elliptical orbit was the first of three laws of planetary motion Kepler discovered.

Kepler also found that a line between a planet and the Sun swept at a regular rate as the planet orbited. And his third discovery simply entranced him.

It said that the square of a planet's orbital period (in years) equals the cube of its average distance from the Sun (in units of the Earth–Sun distance). For astronomers and mathematicians, a simpler and more powerful harmony would be hard to find.

GALILEO'S NEW WORLDS

Another Copernican, Galileo Galilei (1564–1642) of Italy became in 1609 the first person to use a telescope to investigate the heavens (see p. 34). His discoveries were nothing short of spectacular: craters and mountains on the Moon, the four largest satellites of Jupiter, sunspots, the phases of Venus, and the starry nature of the Milky Way. In March 1610, Galileo published *The Starry Messenger*, detailing his initial observations using a telescope. It astounded astronomers everywhere.

But Galileo's discoveries conflicted with the Church's Earth-centered view of the universe because they fit much better with the Copernican

TABVLA III. ORBIVM PLANETARVM DIMENSIONES, ET DISTANTIAS PER QVINQVE REGVLARIA CORPORA GEOMETRICA EXHIBENS.

ILLVSTRISS: PRINCIPI AC DNO, DNO, FRIDERICO, DVCI WIR: TENBERGICO, ET TECCIO, COMITI MONTIS BELGARVM, ETC. CONSECRATA.

295. Kepler, 1596. (Greatly reduced.)

model than the Ptolemaic. Some Church officials denied the existence of Jupiter's moons—and even refused to look through a telescope. At a trial in Rome in 1633 Galileo was found guilty for having "held and taught" the Copernican doctrine, and was forced to renounce his beliefs. He spent the last nine years of his life under house arrest.

HEAVENLY CLOCKWORK

In the same year that Galileo died, Isaac Newton was born in England. Newton (1642–1727) was to put the astronomy of Copernicus and Kepler on a solid mathematical footing.

Newton developed mathematical tools that could analyze planetary motions, and these led him to develop a theory of universal gravitation. The action of gravity, Newton saw, underlay Kepler's laws of planetary motion. Further, Newton showed how, with simple modifications, the same forces at work in the universe could be seen controlling objects in the everyday world.

Newton set down his ideas in *The Mathematical Principles of Natural Philosophy*, better known by a short form of its Latin title, the *Principia*. Published in 1687, the book laid the cornerstone for all physical science down to the 20th century. Only with relativity theory in the early 1900s did physicists and astronomers encounter the limitations of the Newtonian universe.

Newton's grand system can be likened to a vast machine, rolling along with a clocklike regularity

that can be understood and predicted. The mathematical rigor of the *Principia* put to rest the last fragments of antiquity's astronomy —and opened an infinite universe for discovery.

Starry Discussion
Galileo's *Dialog on the Two Chief World Systems* (1632) set down his Copernican views. The frontispiece depicts two of the book's characters debating the virtues of Ptolemy and Copernicus, while a third looks on. The book was banned by the Church.

Newton's Achievement
Isaac Newton discovered the laws of gravity and solved the problem of planetary motion.

An Expanding
Universe

Over the last few centuries, great leaps in our knowledge of the universe followed great leaps in technology and theory. Bigger and better telescopes were joined by new instruments that could "see" invisible light. Grand theories explained unanswered questions. And manned space missions and robot probes mapped the Solar System in detail.

"Space isn't remote at all. It's only an hour's drive away if your car could go straight upwards."

Fred Hoyle
British astronomer, b. 1915

Astronomy Comes of Age

Until the 1800s, astronomers had little idea what celestial objects were made of. To go beyond mere position measurements, astronomy had to embark on a shotgun marriage with physics.

Herschel's Giant
By the end of the 18th century, astronomers were observing the skies using bigger and better telescopes. One of the biggest was William Herschel's 48 inch (1.2 m) reflector at Slough, England.

At Isaac Newton's death in 1727, astronomers were left with a clockwork model of the universe. Precisely on schedule, moons orbited planets and planets orbited the Sun while comets came and went. In the background the unchanging stars shone upon this great machine, and provided an apparently fixed frame of reference. Astronomers used improved telescopes and other instruments to measure celestial motions in ever-finer detail, filling out the grand scheme that Newton had formulated.

There were surprises, however. In March 1781, the English astronomer William Herschel (1738–1822) discovered the planet Uranus, the first new planet since antiquity. But discovering planets was not Herschel's main goal. He was systematically sweeping the skies, looking for objects beyond the Solar System—and finding a lot of them. Yet few astronomers followed Herschel's path, and in fact the next planet to be found (Neptune, in 1846) followed from a careful use of Newtonian celestial mechanics, the rules governing the clockwork's action. It seemed as if astronomy would forever remain ignorant of what stars and planets were, even as it told with impressive accuracy how they moved.

LIGHT'S FINGERPRINTS

A change began in 1814, when the German physicist Joseph Fraunhofer (1787–1826) found dark lines crossing the rainbow spectrum of the Sun.

He had little idea what they were, and an explanation emerged only in the 1850s, when two other German physicists, Gustav Kirchoff (1824–87) and Robert Bunsen (1811–99), invented the spectroscope. This device breaks up light and lets scientists measure its colors with high precision. Kirchoff and Bunsen found that each chemical element and compound put distinctive lines into a spectrum, and with this breakthrough astronomers could

---

HOW FAR THE STARS?
For measuring distances between the stars, miles and kilometers are poor units to use—like measuring a crosscountry trip in hair's-breadths. A more useful unit is the *light-year*, the distance light travels in one year, or about 6 trillion miles (10 trillion km). Another unit is the *parsec*, equal to 3.26 light-years.

Within the Solar System, however, such units are far too large, while kilometers and miles remain awkwardly small. For interplanetary distances, scientists often use the *astronomical unit* (AU), the average distance between Earth and the Sun. This is 93 million miles (150 million km). In this scheme, Mercury orbits at 0.39 AU, Venus at 0.72, Mars at 1.52, Jupiter at 5.2, and so on.

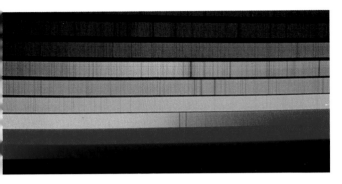

Fraunhofer Lines
In this spectrum of the Sun's light, the dark Fraunhofer lines can be clearly seen. Each line is produced by atoms in the Sun's atmosphere that absorb light at specific wavelengths. By studying such lines, astrophysicists can determine what a star is made of, how hot and how bright it is, and many other properties.

finally discover what stars were made of. (Mostly hydrogen, it turned out.)

The spectroscope proved the key to identifying the mysterious "nebulas"—softly glowing clouds of light—that Herschel and others had found in such great numbers scattered through space. Analysis of spectra uncovered big differences between nebulas—some showed only the bright emission lines of one or two elements and were clearly clouds of thin gas, while others were more Sun-like. The first type really are nebulas, and many of them are places where new stars are forming. But the clouds with stellar spectra are now called galaxies and are known to be vast collections of stars.

The spectroscope also let astronomers measure the speed of a star or nebula toward or away from us by noting a change in the wavelength of spectral lines. This is called the Doppler effect, after its discoverer, the Austrian physicist Christian Doppler (1803–53).

Finally, spectroscopy uncovered entirely new chemical elements, including helium, discovered

in the Sun's spectrum in 1868. It was not found on Earth for another 30 years.

The spectroscope caused only part of astronomy's changes during the 1800s. Later in the century, photography began replacing the human eye at the telescope, and around 1900 astronomers were using the first crude electronic detectors to measure the brightness of stars.

PAPER STARS

At about this time, an entirely new kind of astronomer appeared on the scene. The astrophysicist combined theoretical physics with observations.

The German physicist Max Planck (1858–1947) developed quantum theory in 1900. Five years later, Albert Einstein

Doppler Shift
The direction of movement of stars and galaxies can be detected by analyzing the light from them and using the principle of the Doppler shift. If a light source stays motionless with respect to Earth, the lines in its spectrum appear in the "standard" position (1). If the source is moving away from Earth, the light waves are longer and the lines shift toward the red (2). If the object is moving toward Earth, the waves are shorter and the lines shift toward the blue (3).

1 No shift = galaxy at rest

2 Shift to red = galaxy moving away from Earth

3 Shift to blue = galaxy approaching Earth

Einstein's Revolution
Albert Einstein's complex theories rewrote the laws of physics which had been unchallenged since Isaac Newtons's time.

(1879–1955) published his theory of special relativity, following it in 1916 with his theory of general relativity. Taken together, these discoveries linked space, time, and velocity, creating completely new ways of imagining the universe. In this strange synthesis, light beams bend under the force of gravity, clocks run at different rates for different observers, and the universe expands endlessly.

Picturing the Stars
These days, most professional observations using optical telescopes are carried out with photographic or electronic equipment. The trend had its beginnings in the late 19th cenutry. This photo of the Whirlpool Galaxy was taken in 1910.

In only a few years, astronomy had journeyed very far from Newton's tidy scheme.

By the beginning of the 1930s, astrophysicists were at work building so-called "paper stars"— theoretical models of how stars form, live, and die. In Britain, James Jeans (1877–1946) and Arthur Eddington (1882–1944) used the new science of nuclear physics to explain how stars shine. In the United States, Subrahmanyan Chandrasekhar (1910–1995) wrote the first mathematical formulations of stellar processes—including how stars might explode and how white dwarfs might form.

Theoretical astronomers' models grew in complexity after the Second World War, when electronic computers gave them a powerful new tool for simulating how stars and even galaxies work. Theoretical models could then be put to the test against real-world observations made using increasingly sensitive instruments.

GEORGE HALE'S GLASS GIANTS

The early 1900s saw an explosive growth in the size of telescopes and in their light-grasp, much of it due to a single man, George Ellery Hale (1868–1938), a U.S. astronomer. Hale used his ample powers of persuasion to convince industrial tycoons and wealthy foundations to fund the world's largest telescope—and not once, but four times in a row.

The first of Hale's "glass giants" was the 40 inch (1 m) refractor at Yerkes Observatory in Wisconsin (1897), still the largest of its type. But Wisconsin skies are often cloudy, and Hale next settled upon Mount Wilson in the San Gabriel Mountains above Los Angeles, then smog-free. There he constructed several solar telescopes (Hale's own scientific specialty), plus a 60 inch (1.5 m) reflector (1908); a 100 inch (2.5 m) reflector followed in 1917. It was the 100 inch that Edwin Hubble used to make many of his ground-breaking observations of galaxies (see p. 46).

But as Los Angeles skies grew brighter during the 1920s, Hale once more looked around, and found the sparsely inhabited dark-sky country between Los Angeles and San Diego. On Palomar Mountain, he chose the site for a 200 inch (5 m) telescope. Work began in 1928 but progressed slowly as the builders confronted one engineering hurdle after another. Then the Second World War intervened. Finally, in 1948, 10 years after its builder's death, the great telescope and its protective dome were complete (bottom). It bears Hale's name to honor a lifetime's contribution to astronomy.

For several decades, the Hale was the world's largest telescope. Recent instruments surpass it in size, but retro-fitting with new detectors has kept the 200 inch at the forefront.

Galaxy Man

Edwin Hubble's discoveries overturned old ideas about the nature of the universe.

Nebulas or Galaxies?

In 1923, Edwin Hubble measured the distance to the Great "Nebula" in Andromeda (M 31) using the new 100 inch (2.5 m) telescope at Mount Wilson. It was so far away, he found, that it could not be part of our Galaxy, as previously believed. It was, in fact, a galaxy like ours.

NEW REALMS

Telescopes grew greatly in size and quality through the early decades of the 20th century, especially in the U.S. (see box, p. 45). All astronomers benefited from the new giants, but cosmologists gained most. These specialists study the origin and development of the universe, and focus mainly on faint, distant galaxies—the fainter the better.

In the 1920s, the American Edwin Hubble (1889–1953) measured the distance to the Andromeda Galaxy and was the first to put an accurate yardstick on the universe. Using the largest telescopes, he went on to photograph thousands of galaxies. His results included a classification of galaxies which, in modified form, is still in use. More important was his discovery that the fainter and smaller a galaxy appears, the more its spectral lines are shifted toward red colors and the faster it is receding from us. This showed that the universe is expanding, a prediction that Einstein's relativity theory had made.

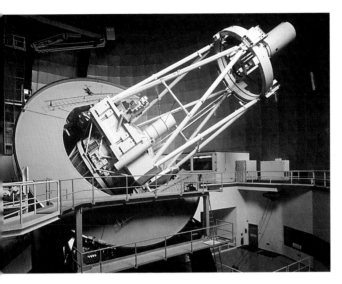

Going Electronic
The Anglo-Australian Telescope at Siding Spring, Australia, has a 154 inch (3.9 m) mirror. Completed in 1975, it was one of the first telescopes able to be controlled electronically.

EVEN BIGGER TELESCOPES

Astronomy's need for more light—and bigger telescopes—is inescapable because that is all the universe sends us.

Following Hubble's lead, cosmologists today use even larger telescopes to map the expanding universe in ever greater detail. The two 400 inch (10 m) Keck telescopes operate on Mauna Kea in Hawaii, and numerous instruments span the gap between these and Mount Palomar's 200 inch (5 m). Even bigger telescopes are planned, some destined to go into orbit where they will escape the distorting effects of Earth's restless atmosphere.

Today's glass giants are compact and highly automated. Computers run observatories: they aim the telescopes, operate the instruments, calibrate the data, and keep the optics in line. And computer-controlled adaptive optics cancel out much atmospheric distortion and produce sharper images.

Largest Scopes
Large telescopes are prodigious light-gatherers, and the largest of all are the 400 inch (10 m) Keck I and Keck II telescopes on the summit of Mauna Kea, Hawaii. Sited to take advantage of the clear, steady air, and collecting far more light than smaller instruments, they see the heavens in great detail.

Beyond the Visible

The visible heavens are just a tiny part of the universe. To see the whole thing as it really is astronomers tap into invisible wavelengths and fly their instruments far away from Earth.

Very Large Array
The Very Large Array (VLA) radio telescope consists of 27 movable dish antennas arranged in a Y-pattern in the New Mexico desert. Each arm of the Y is 13 miles (21 km) long. The antennas can be spread for maxiumum resolving power or clustered (as here) for greater sensitivity.

A rainbow's gentle glow scarcely hints at the wonders wrapped in the word "light." On a dark night we see a universe full of stars, nebulas, and galaxies. Yet all that we see, even with the most powerful optical telescope, is but a small fraction of what the universe contains. For ages astronomy's picture of the heavens was painted solely in terms of what the human eye could see. Then in the 1800s, physicists discovered

that visible light was just one part of the full array of energy nature produces. The entire range is called the electromagnetic spectrum, and it encompasses all kinds of radiation, from high-energy gamma rays to low-energy radio waves. In the middle, like a door opened just a crack, lies the realm of visible light.

Astronomy was very much a visual science until the 20th century when astronomers began

RADIO WAVES   **INFRARED**

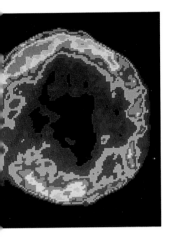

to realize that they could not continue to peer just through light's narrow crack. To truly understand how the universe works in all its complexity, all forms of radiation are fair game. Special telescopes are required, and some of them must be placed in space to avoid the filtering and distortion caused by Earth's atmosphere.

THE LONG OF IT...

Astronomers' first step outside the visible part of the spectrum was the infrared, where wavelengths are just longer than visible light. William Herschel first detected infrared (heat) radiation in 1800, but his discovery outran the physics of his time. And for lack of the proper technology, infrared astronomy remained a curiosity

until the 1960s. Today, infrared astronomy is vital for understanding how stars form, since infrared light can reveal what is happening in the depths of gas and dust clouds where stars are born. The Space Infrared Telescope Facility (SIRTF) belongs to a suite of space-borne NASA telescopes called the Great Observatories. Others include the Hubble Space Telescope (see p. 52) and the Chandra X-Ray Telescope (see pp. 50–51).

Radio is a form of light with wavelengths longer than infrared. In the late 1800s, astronomers began looking for radio waves predicted by theory, but again technology fell short. In 1932 Karl Jansky, an American telephone engineer, investigated static heard in long-distance phone calls and found it was *long* distance indeed: radio noise from the Milky Way Galaxy! From the early 1950s, research activity increased greatly and the radio universe opened up. Radio astronomers discovered both pulsars, the spinning remnants of dying stars, and quasars, compact objects thought to be the active cores of very distant galaxies. They have also observed the cosmic background radiation, the fading glow of the Big Bang that formed the universe.

Radio Remnant
Tycho's supernova remnant is an expanding shell of gas produced when a star exploded as a supernova in 1572. A radio image shows that the hottest parts of the gas shell are around the rim, color-coded yellow and red.

Across the Spectrum
Our Sun produces light most strongly in one small part of the electromagnetic spectrum, which the human eye has evolved to make use of. But the universe works with a much broader palette of "colors." These range from short-wavelength gamma rays and X rays, through the visible, and down into the long-wavelength infrared, microwave, and radio segments of the spectrum. By developing specialized instruments to detect this non-visual radiation, we are able to sidestep the limitations of eyesight to see the universe as it truly is.

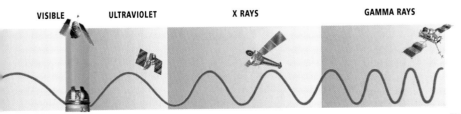

VISIBLE   **ULTRAVIOLET**   **X RAYS**   **GAMMA RAYS**

DISCOVERING SCORPIUS X-1
The first cosmic X rays were picked up in 1962 by Geiger counters aboard a small U.S. rocket. The mission's intention had been to detect X rays coming from the Moon. But when the rocket returned to Earth, scientists were surprised to find no lunar X rays at all. What the instruments *had* seen were X rays coming from an unassuming star in the constellation Scorpius. Christened Scorpius X-1, this binary-star system is the brightest X-ray source in the sky.

Ultraviolet Sun
Observing the Sun at ultraviolet wavelengths highlights many details of the high-temperature solar corona, including sunspots, solar flares, and coronal mass ejections. The image was taken by the Solar and Heliospheric Observatory (SOHO), launched in 1995 by NASA and the European Space Agency.

Most radio telescopes are large metal parabolic dishes, which focus the faint signals they receive from space. Big radio telescopes include the 250 foot (76 m) one at Jodrell Bank, England, and the 1,000 foot (300 m) antenna at Arecibo, Puerto Rico.

To gain sharper views, astronomers link a number of radio telescopes together, creating, in effect, one giant-size telescope. Examples include the Very Large Array in New Mexico (27 antennas spread across many miles) and the Very Long Baseline Array, with 10 antennas from Hawaii to Puerto Rico.

...AND THE SHORT

Radio waves penetrate the atmosphere relatively easily, but light at wavelengths shorter than the visible is mostly absorbed. Astronomers interested in studying ultraviolet radiation, X rays, and gamma rays have to lift their telescopes above the atmosphere.

Ultraviolet light lets astronomers investigate very hot objects in the universe, such as gas around ordinary stars and hot stars whose evolution is running faster than the Sun's. This is the task of NASA's Far Ultraviolet Spectroscopic Explorer (FUSE).

Shorter than the ultraviolet are X rays. Because they are such an energetic source of radiation, X rays allow astronomers to study violent and extremely energetic objects and processes, such as supernova explosions. X rays are also produced when charged particles interact with strong magnetic fields, a common process in the depths of galaxies. NASA's Chandra X-Ray Observatory has discovered a new kind of medium-size black hole, unsuspected by theory, and has seen X rays coming from comets.

Gamma rays are the most energetic form of radiation known. They slice through ordinary optics, so they cannot be observed directly. Instead, special detectors are monitored for evidence of changes brought about by these rays. Gamma rays provide a look at the oddest, most exotic objects in the universe, such as black holes, the swirling centers of active galaxies, and the Sun's hottest regions.

A satellite observatory which mapped this high-energy form of radiation was the Compton Gamma-Ray Observatory (CGRO), brought down from orbit in 2000 when its main systems began to fail. It will be replaced by the Gamma-Ray Large Space Telescope (GLAST), planned for launch in 2005.

The past century has taught astronomers to see with new eyes, and "invisible astronomy" has deepened our understanding of the universe and how it works. For all the wonders of the cosmos that greeted our great-great grandparents, the universe we are now exploring is vastly greater, stranger, and more interesting.

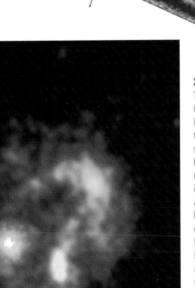

X-ray Vision
The Chandra X-ray Observatory (above), launched in 1999, tunes in to high-energy regions of space, including black holes and supernova remnants. It captured this image (left) of Eta Carinae, an unstable star that may become a supernova. The outer yellow-orange, horseshoe-shaped ring has a temperature of more than 5 million°F (3 million°C). The blue core is much hotter, reaching tens of millions of degrees.

Hubble's All-seeing Eye in Space

With a mirror 94 inches (2.4 m) across, the Hubble Space Telescope (HST) is not the largest telescope in existence, but it has the best view of all. Orbiting 380 miles (600 km) above Earth, HST has observed all kinds of astronomical bodies, from the Moon to the most distant galaxies.

HST was launched in 1990 and is more than halfway through its planned lifetime. Scientists are now planning the Next Generation Space Telescope, an orbiting giant with a mirror 26 feet (7.9 m) across and a light grasp more than 10 times that of HST.

Maintenance Schedule
A shuttle astronaut on a robotic arm installs protective covers on the Hubble Space Telescope's magnetometers. The HST receives regular maintenance and equipment upgrades from the space shuttle, keeping it at astronomy's leading edge.

What Hubble Has Seen
From the planets and moons of our Solar System and the spiral arms of nearby galaxies, to young galaxies at the edge of the universe and supermassive black holes in quasars—the Hubble Space Telescope has seen it all. By the year 2000, it had looked at 14,000 objects and made more than 330,000 exposures. This selection shows, from top to bottom: Jupiter's Galilean satellites; HH32, a young star ejecting jets of gas into space; and a binary-star system.

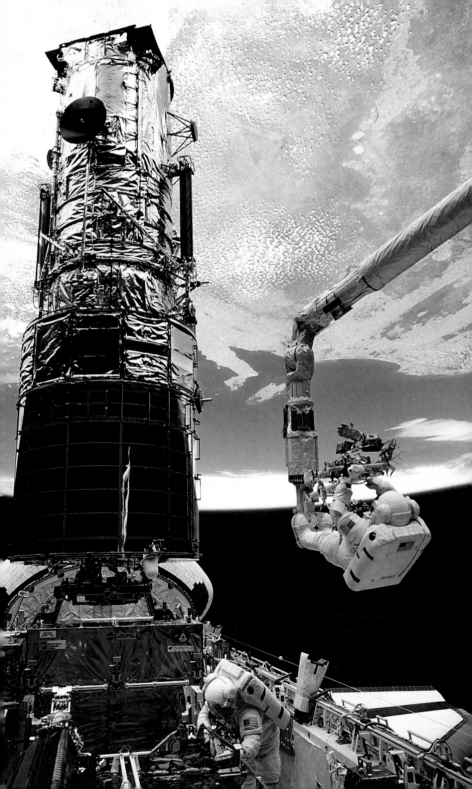

Stepping into Space

Humans began to venture into space ages ago, if only in works of fantasy and science-fiction. It took a superpower rivalry to make it happen for real.

For countless years, people have looked at the Moon, planets, and stars, and wondered about traveling there. In antiquity, Lucian of Samos wrote a romance about a hero whose boat is carried to the Moon by a waterspout, and who later returns to the Moon on a pair of self-made wings. In 1634, Johannes Kepler wrote a tale about going to the Moon by magic. By the 19th century, astronomers knew much more about space, so when Jules Verne wrote *From the Earth to the Moon*, he mentioned weightlessness and the need to carry air.

REAL SPACEFLIGHT

Such stories launched many dreams, yet few payloads. In 1919, the U.S. physicist Robert Goddard proposed using rockets to lift instruments to very high altitudes—even to the Moon. During the 1920s and 30s, he built and tested several liquid-fuel rockets, one reaching 8,000 feet (2,400 m). At about the same time, German engineers were also

FROM V-2 TO SATURN V
Among a talented group of German rocket engineers in the 1930s was Wernher von Braun, a spaceflight enthusiast since childhood. The group soon caught the eye of the military, rearming under the new Nazi regime. During the Second World War, von Braun's group created a ballistic missile— the V-2—which could drop an explosive warhead on England.

After Germany was defeated in 1945, the U.S. took over much of the Germans' technical know-how, bringing von Braun and many of his team into its rocketry program. Experiments with captured V-2s (right) led to vast improvements in rocket design, culminating in the 1960s with the Saturn V rocket, which put man on the Moon.

developing rockets. Their work had a lasting effect on spaceflight history (see box, facing page).

THE COLD WAR STARTS A SPACE RACE

The late 1940s and early 50s marked the beginnings of the Cold War between the U.S. and the Soviet Union. In this period of mistrust, competition, and military build-up, scientists on both sides worked in secret to produce rockets capable of carrying nuclear warheads. The first spaceflights evolved from this military research and they, in turn, became part of the superpower rivalry.

The Soviet Union took an early lead in what became popularly known as the space race when they launched the first artificial satellite, Sputnik 1, in October 1957. A month later they launched Sputnik 2, an even larger satellite with a live dog inside. The first U.S. satellite, Explorer 1, did not fly until January 1958.

Human spaceflight was the next goal, and once again, the Soviets were first, as Yuri Gagarin soared for one orbit of Earth in

Canine Cargo

In 1957, Laika the dog became the first living thing in space when the Soviet Union sent Sputnik 2 into orbit. The craft was not designed to return, and the animal was put to sleep before the life-support system gave out.

Gagarin's Triumph

The Soviet Union took the lead in the space race when Yuri Gagarin (left), a Soviet Air Force lieutenant, blasted off on April 12, 1961, becoming the first person to fly in space. He completed a single orbit before returning to Earth in the descent module of his Vostok 1 spacecraft (above). The whole flight lasted 108 minutes from launch to landing.

Woman in Space
The first woman in space was the cosmonaut Valentina Tereshkova. In June 1963, she orbited Earth 48 times aboard Vostok 6.

April 1961. Three weeks later, the U.S. sent Alan Shepherd on a flight that left the atmosphere but did not go into orbit. The first American to orbit Earth was John Glenn, who made three orbits in February 1962. By that time, the Soviets had made a second flight which stayed up a day and made 17 orbits.

The U.S., however, was doing better than it appeared—the American Mercury spacecraft was more sophisticated, being under the astronaut's full command, whereas the Soviet Vostok craft was flown by an onboard autopilot and ground control.

A Walk in Space
On June 3, 1965, Edward White opened the hatch of his Gemini 4 capsule and took a walk in space, the first American to do so. Tragically, White died 18 months later, along with two fellow crew members, when a fire destroyed their Apollo spacecraft on the launchpad.

KENNEDY SETS THE TARGET: THE MOON

The space race kicked into high gear when President John F. Kennedy announced in May 1961 that before 1970 the U.S. would send a crew to the Moon and return them to Earth. For their part, the Soviets announced

no goals, but nobody doubted the Moon was their target also.

To achieve the Moon landing, the U.S. laid out a complex plan. After Mercury, a two-man Gemini craft would test techniques of joining spacecraft in orbit. It would also remain aloft for two weeks, the length of a lunar trip. Following Gemini, a three-man Apollo spacecraft, plus a gigantic booster rocket, would be developed to go to the Moon.

The Soviets modified Vostok to carry a crew of three, and also sent up the first woman cosmonaut, Valentina Tereshkova, in June 1963. In March 1965, Alexei Leonov became the first to walk in space, drifting near his spacecraft on a tether. In June 1965, Edward White of the U.S. did the same from a Gemini craft. As 1966 ended, the Gemini program showed that U.S. spacecraft could change their orbits, which the Soviet craft could not. And Apollo's booster rocket, the Saturn V, was nearing flight tests while the Soviet lunar booster was still being planned.

THE PRICE OF SPACEFLIGHT

In 1967 the space race started to cost more than just money. In January, a fire flashed through an Apollo spacecraft on the ground,

killing all three astronauts. In April, the Soviet Union lost a cosmonaut when his Soyuz spacecraft crashed on landing. Both countries grounded all flights while engineers studied the failures.

In December 1968, the Apollo 8 spacecraft was launched by Saturn V rocket, and successfully circled the Moon for a day and returned to Earth. In March 1969, Apollo 9 tested the lunar lander in Earth orbit. This spidery craft was to take two astronauts from lunar orbit to the surface and bring them back to lunar orbit to rejoin the command module for return to Earth. In May 1969, Apollo 10 flew a full dress-rehearsal, flying the lander to within 50,000 feet (about 15,000 m) of the Moon's surface

ONE GIANT LEAP...

As the Apollo program neared its high point, the Soviet lunar program reached a new low. In early July 1969, the Soviets' lunar booster exploded on the launch pad. Ten days later, they launched an unmanned spacecraft to the Moon, but the probe crashed.

On July 20, 1969, Apollo 11 reached lunar orbit and Neil Armstrong and Edwin Aldrin

touched down on the Sea of Tranquillity. Armstrong said, "That's one small step for a man, one giant leap for mankind." Aldrin and Armstrong spent two hours on the surface collecting rock samples and setting up experiments. Their safe return to Earth marked both the end of the space race and a great scientific triumph.

Apollo 17 brought an end to the program in December 1972, by which time a dozen humans had walked on the Moon and over 1,000 pounds (450 kg) of lunar rocks had been returned. If the space race rode on a wave of

Mission Accomplished
Astronaut James Irwin salutes the flag in front of Apollo 15's lander and lunar rover. The Apollo program was perhaps the most elaborate technological and scientific endeavor in human history. Born during the Cold War, it managed to transcend politics in the end, by gripping the imagination of millions of people across the world.

MEN ON THE MOON		
Apollo 11, July 20, 1969	Neil Armstrong and Edwin Aldrin	*Mare Tranquillitatis*
Apollo 12, November 19, 1969	Charles Conrad and Alan Bean	*near Mare Cognitum*
Apollo 14, February 5, 1971	Alan Shepard and Edgar Mitchell	*near Fra Moro crater*
Apollo 15, July 30, 1971	David Scott and James Irwin	*near Montes Apenninus*
Apollo 16, April 20, 1972	John Young and Charles Duke	*Descartes region*
Apollo 17, December 11, 1972	Eugene Cernan and Harrison Schmitt	*Montes Taurus*

A Journey Begins

The space shuttle Atlantis is launched into the skies above Cape Canaveral. Each of the shuttle's three main engines has enough thrust to power two and a half jumbo jets. In contrast to this explosive beginning, at the end of its mission the shuttle will glide, unpowered, to the ground.

political, military, and techno-logical competition, it nonetheless put humankind on the long road to the stars.

SHUTTLES AND STATIONS

The U.S. space program shifted toward developing a reusable spacecraft, the space shuttle, which would not venture far from Earth. Columbia, the first shuttle, flew in April 1981. Three more shuttles were built and all flew routinely, until January 1986 when Challenger exploded soon

after launch, killing its crew. The fleet resumed activity in 1988, launching probes to Venus and Jupiter, and putting the Hubble Space Telescope into orbit.

Designed as a space "truck," the shuttle can carry up to 30 tons (27 tonnes) and a crew of eight for a two-week mission. Its pay-loads include scientific instrument packages for conducting experiments impossible on Earth.

The shuttle has been used in the construction of the International Space Station, a project

Weightless in Space
As these space shuttle astronauts show, weightlessness can be a source of great enjoyment. But it can also cause problems, including nausea and loss of muscle tone.

using components from many countries, including Russia, Japan, France, and Canada. (Earlier space stations, such as Russia's Mir and the American Skylab, paved the way by testing technologies.)

Orbital assembly began in 1998, with new modules added over following years. The first inhabitants—an American and two Russians—took up residence in the partly built station in November 2000. A succession of crews should complete construction by 2005. The station can house up to seven people, who will conduct experiments in manufacturing, engineering, and technology. The station will also help scientists gain biomedical knowledge for manned expeditions into the Solar System.

International Effort
When complete, the International Space Station will measure a spacious 356 feet by 290 feet (110 by 90 m), large enough for as many as seven crew and their equipment. It will receive regular deliveries by space shuttle.

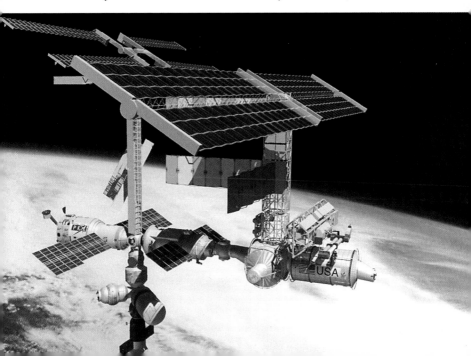

Bouncing Down for a Sojourn on Mars

In July 1997, the Mars Pathfinder spacecraft landed on the red planet at Ares Vallis, a dusty, rocky valley that had once been a torrent of water. With its cushioning airbags deflated, Pathfinder set about its major task: the safe delivery of Sojourner, a six-wheeled rover. Steered from Earth, Sojourner ran down a ramp from the mother craft and roamed the area nearby, testing the makeup of rocks with a battery of instruments. By the time its power system failed after almost a hundred Earth days, Sojourner had taken thousands of photos and geological measurements, greatly increasing our knowledge of Mars.

Pathfinder lander separates from orbiter

Parachute opens, slowing the lander's descent

Airbags around lander inflate

Making Tracks

Soon after reaching the surface, a camera on the Pathfinder lander pictures the six-wheeled Sojourner and the remains of the airbags (upper photograph). A rocky red surface stretches away to low hills in the distance. Next, Sojourner emerges and begins its work. In the lower picture, it is heading toward a large rock and will use its spectrometer to gather geological data. Sojourner found evidence for rock types previously unsuspected on Mars.

Landing by Airbags

The Mars Pathfinder mission tested a new way to land a spacecraft—airbags. Unlike earlier landers, which were lowered by rockets, airbags do not contaminate rock samples with rocket exhaust. The new method worked perfectly.

Lander cocooned in airbags reaches surface

Airbags deflate

Airbags retract and petals of lander open

Sojourner rover leaves lander and explores nearby area

Probing the Solar System

Astronauts and cosmonauts are rightly feted for their courage and skill in a hostile environment. But could it be that the real heroes of space exploration are the far-flying robots that have opened up the Solar System?

Venera
Between 1970 and 1982, 10 Soviet Union Venera landers reached the surface of Venus. The immense heat and pressure made each lander malfunction after a few minutes, but four of the landers managed to photograph their surroundings, showing gritty sand and slabs of volcanic rock.

Human spaceflight is undeniably dramatic. Yet the most far-reaching discoveries about the planets have come from the data radioed back to Earth from unmanned spacecraft roaming the Solar System. Following a time-tested sequence—flybys first, then orbiters, then landers— scientists have used these robot probes to steadily chip away at the ignorance surrounding Earth's neighbors.

VISITS TO HOT SPOTS
Closest to the Sun, the planet Mercury has seen just one spacecraft, the United States' Mariner 10, which made three flyby visits in 1974–75. The images and data it sent back to Earth revealed an airless Moon-like world of craters and rolling lava plains, broiling under a Sun

that makes the surface so hot (800°F, 430°C) that lead would flow like water. Mariner also found a large iron core hinting that today's Mercury might be the remnant of a once-larger planet that was struck by another body early in its history. In 2008, a U.S. mission called Messenger will return to Mercury to complete the survey. Messenger will also see if ice exists in permanently shadowed craters near the planet's poles, suspected from Earth-based radar studies.

Next out from Mercury is Venus, which has drawn nearly two dozen missions from the U.S. and the old Soviet Union (now Russia). Mariner 2 (1962) found that Venus' surface was as hot as Mercury and crushed under an atmosphere nearly a hundred times heavier than Earth's. The

SOHO's Sun
The SOHO probe took this ultraviolet image of the Sun and its outer atmosphere, the corona. The solar wind that blows past Earth comes from the outermost part of the corona (outside the black circle).

Soviet Venera probes included four landers that radioed back photos of a surface strewn with lava rocks and gravel. To study the planet's features, radar was used to penetrate the cloud deck from orbit. Veneras 15 and 16 (both 1983) and the U.S. Pioneer Venus (1978–92) and Magellan (1990–94) missions mapped Venus' volcanic landscape in great detail.

ROBOTS SCOUT OUT THE RED PLANET

Mars has always seemed the most Earth-like planet, but when the U.S. Mariner 4 flew past in 1965, it photographed a cratered landscape that resembled Earth's Moon. Later, Mariners 6, 7 (both 1969), and 9 (1971) portrayed a somewhat different Mars—well-cratered, but also having water-carved channels, giant volcanoes,

Mercury by Mariner
The first close-up views of Mercury were provided by Mariner 10 in 1974–75. It photographed about 45 percent of the planet, revealing a landscape bruised and battered by thousands of impacts.

TAKING AIM AT THE SUN

Numerous missions have focused on the Sun. These include the joint U.S. and European Solar and Heliospheric Observatory (SOHO), launched in 1995, which has studied the Sun's corona and activity from space. Ulysses (1990–), also U.S. and European, passes over the Sun's poles every five years to view from directions we cannot see from Earth. NASA is planning a mission called Solar Probe, which will pass within three solar radii of the Sun's surface, peering out from behind an *extremely* good heat shield and collecting data from close range.

and deep valleys. Hoping to find life, NASA scientists sent Viking 1 and 2 (1976), two pairs of orbiter and lander spacecraft. While the landers looked for chemical signs of life, the orbiters took tens of thousands of photos, mapping the geology of Mars. No evidence of life was found.

Following up on Viking, NASA's Mars Pathfinder lander (see p. 60) and the Mars Global Surveyor orbiter arrived in 1997. Pathfinder carried a rover which drove around the lander, examining rocks. Global Surveyor's highly detailed images are re-writing the Mars geology book, as they point to a Mars that is warmer, wetter, more volcanic, and geologically younger than scientists had thought.

NASA's Mars plans include a pair of rovers that will land in 2004. Later in the decade, possibl sample-return missions may dovetail with the European Spac Agency's Mars Express orbiter

Mars by Viking
One of the most successful and scientifically productive encounters with a planet was the Viking mission. While two orbiters made a complete photographic survey of the red planet (this picture is a mosaic of Viking images), their landers touched down on the planet's surface to search for traces of life.

Europa by Galileo
The astronomer Galileo first saw Europa as a bright pinprick, along with three other Jovian moons, through a telescope in 1610. In 1995, a space probe named after him revealed Europa's icy surface in unprecedented detail. These two images show differences between fine-grained ice (light blue), coarse-grained ice (dark blue), and ice containing rocky material (brown).

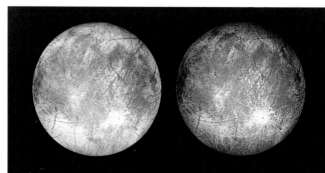

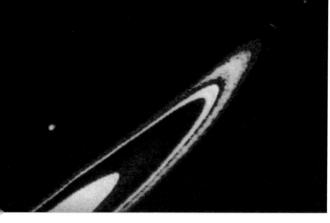

Saturn by Pioneer 11
Pioneers 10 and 11 were the first spacecraft to investigate the outer Solar System. This view of Saturn's rings was beamed from Pioneer 11 in 1979.

nd Beagle 2 lander. The goal of ll this exploration is an ambitious one: to look for evidence of life.

EXPLORING MICROWORLDS

Millions of asteroids, or minor planets, orbit between Mars and upiter. Only a few have been visited. Bound for Jupiter, NASA's Galileo probe flew past Gaspra 1991) and Ida (1993), discovering that the latter has a tiny moon, Dactyl. The Near-Earth Asteroid Rendezvous (NEAR) raft visited Mathilde in 1997 on ts way to orbiting Eros in 1999.

A planned Japanese mission called MUSES-C will visit steroid 1998 SF36. The flightplan alls for MUSES-C to orbit the steroid and pick up surface amples, returning them to Earth n 2007. Samples are highly rized because spacecraft have nown that each of these microworlds differs markedly in density, ge, extent of cratering, and hemical composition.

CLOSE ENCOUNTERS WITH THE GAS GIANTS

Beyond the asteroid belt lie the as giants. Flight times to these istant planets are measured in

years, rather than in months as for the inner planets. Into this daunting realm only U.S. spacecraft have ventured: Pioneers 10 and 11, Voyagers 1 and 2, Galileo, and Cassini.

Pioneer 10 made the first detailed portrait of Jupiter in a 1973 flyby, measuring the planet's powerful magnetic field and imaging its turbulent atmosphere. Pioneer 11 followed the year after, sharpening the picture, and then headed on to Saturn. The much more sophisticated Voyager probes flew past Jupiter in 1979, sending back fascinating images of Jupiter's atmosphere and its varied family of moons. Voyager 2 also discovered that Jupiter has a thin ring.

Reaching Saturn in 1979, Pioneer 11 discovered new rings and drew the first close-up portrait of the Saturnian realm. Voyagers 1 and 2 arrived in 1980 and 1981. As at Jupiter, they detailed Saturn's features, finding new rings and moons and mapping the racing clouds of the planet's atmosphere.

After Saturn, Voyager 1 and both Pioneer probes headed out of the Solar System, while

Decoding Data
Information from space probes is transmitted back to Earth in digital form and has to be processed by computer to be useful. An image from the Voyager 2 flyby of Neptune is being processed here.

Calling Cards
Wearing protective suits, two technicians display "calling cards" from Earth placed aboard both Voyager spacecraft. They include a recording with greetings from world leaders and samples of music, and a gold-plated disk showing Earth in relation to nearby pulsars.

Uranus' Family
Uranus resembles a bright blue cueball in this composite Voyager 2 image of the planet and five of its moons.

Voyager 2 took aim at the little-known worlds of Uranus and Neptune. Arriving at Uranus in 1986, Voyager 2 found a featureless planet shrouded in a blue haze of methane. Next to this bland-seeming world, Uranus' satellites stole the show with their distinctive and complicated geologies.

Voyager 2 completed its tour of the gas giants in 1989, when it reached Neptune. It found an active planet with giant storms in its atmosphere, and frosts and a thin atmosphere on its large moon, Triton. Its main mission completed, Voyager 2 moved out of the Solar System.

NASA returned to Jupiter in 1995, when a spacecraft named Galileo arrived. It shot an entry probe into the Jovian atmosphere and then spent several years surveying the planet and its moons. These include Europa, which has a skin of ice covering an ocean of salty water or slush in which life may be present. (To test this, NASA is planning a Europa Orbiter craft for a closer look.) Another wondrous moon is volcanic Io, where plumes of sulfur shoot hundreds of miles into the sky and lava lakes are hotter than any on Earth.

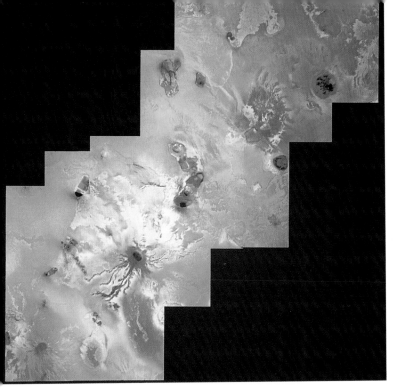

Paralleling Galileo's mission, the Cassini orbiter and probe were launched in 1997 for a rendezvous with Saturn in 2004. Instead of going into Saturn's atmosphere, the Cassini probe is to land on Titan, Saturn's largest moon. Titan appears covered in hydrocarbon deposits and has a thick atmosphere with methane rain and clouds. Scientists think that the chemistry of this moon may resemble that from which life on Earth formed. If so, then Titan may be one of the most likely places to find evidence of extraterrestrial life.

TO THE EDGE

Only Pluto remains unvisited. The U.S. has planned a mission, but high costs make its future uncertain. If flown, it would, after Pluto, continue into the Kuiper Belt of comets lying beyond.

Comets that come closer to Earth are also targets. The Stardust probe is collecting particles from the tail of Comet Wild 2 for return to Earth. In August 2011, the Rosetta mission will study Comet Wirtanen from close range for one Earth year. Rosetta will then land a probe on the comet's surface, and will travel with the comet for another full year.

OUT OF THIS WORLD

Voyager 1 is the most distant human-made object, at more than 7.4 billion miles (11.8 billion km) from Earth. Launched in 1977, it flew by Jupiter and Saturn in 1979–80, and is now nearing the edge of the Solar System. Voyager 1 should keep sending back data until at least 2020.

Volcanic Io

Astronomers suspected that Io, a Jovian moon, might show evidence of volcanic activity, but nothing prepared them for what Voyager 1 found. The images revealed a surface completely covered with volcanic vents and lava lakes—it is the most volcanically active body in the Solar System.

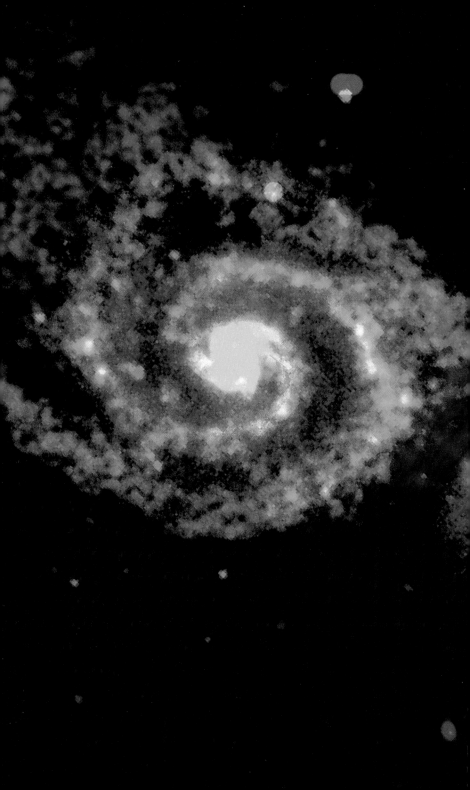

Astronomy and You

Understanding
the Sky

For the novice, the night sky may
at first appear confusing—a clutter
of stars with no real pattern. This
chapter brings order to the chaos
by giving a solid grounding in
some of the skywatching basics.

*"Astronomy compels the soul to
look upward and leads us from this
world to another."*

Plato
Greek philosopher, c. 428–348 BC

Your Place in the Universe

Skywatchers need to know their place: planet Earth orbits the Sun, one of billions of stars in the Milky Way Galaxy, itself one of billions of galaxies in the universe.

Starscape
You have only to look upward on a clear night to appreciate the vastness of space. Of the seven stars that make up the Big Dipper, here glimmering over the Canadian Rockies, the closest is 60 light-years away—that's 360 trillion miles (570 trillion km).

When you gaze at a star-filled sky, you look across vast distances. The Moon, our closest neighbor, lies 240,000 miles (385,000 km) away, far enough that it took Apollo astronauts three days to get there. Find a naked-eye planet, perhaps Jupiter, and you are looking across hundreds of millions of miles—years of travel, even at the fastest rocket speeds.

INTERSTELLAR SPACE
Beyond the realm of the planets lies a gulf in space too large to measure in miles and kilometers. Instead, astronomers speak in terms of light-years—the distance light travels in one year. Alpha Centauri, the nearest bright star beyond the Solar System, lies four light-years away. To visualize that distance, try mentally reducing our Solar System to the size of a suburban backyard. On that scale, Alpha Centauri would be a dot 25 miles (40 km) away across the city. The distance humans have traveled to the Moon would be less than a millimeter.

Our Sun is but one star in the Milky Way, a spiral galaxy of some 400 billion stars that stretches 100,000 light-years from side to side. If the nearest star were

LOOK BACK IN TIME
Everywhere you look into space you peer back into time, to see objects as they used to be. Light from Alpha Centauri, the nearest bright star, takes four years to reach us. In other words, you see it as it appeared four years ago. Find the stars of Orion's Belt and you look back to medieval times. Gaze at the Milky Way, that hazy band of light arcing across the sky, and you stare at light as old as civilization itself. Find the Andromeda Galaxy and you are seeing light emitted when the first hominids wandered the African plains 2.5 million years ago.

indeed across the city, then the Milky Way Galaxy would stretch over 600,000 miles (1 million km) wide, three times the distance to the Moon.

BILLIONS AND BILLIONS

Our Milky Way Galaxy is just one of billions of star-filled galaxies. One of the nearest, the Andromeda Galaxy, can be seen with the unaided eye. Andromeda and the Milky Way are the largest members of a small gathering of at least 40 galaxies called the Local Group.

Our galactic family in turn lives on the out-skirts of a dense cluster of thousands of galaxies called the Coma-Virgo Supercluster. Through a backyard telescope you can see member galaxies of this and many other galaxy clusters whose light is as old as the dinosaurs.

The Hubble Space Telescope has seen galaxies a hundred times more distant still, 10 billion light-years away, so far away that they appear as smudges near the edge of the observable universe. We see these galaxies as they appeared shortly after the birth of the universe.

Our Home

Earth sits third in line from our star, the Sun, among the other rocky planets. Then comes a wide gap to the first of the gas giants, Jupiter. Neptune and Pluto, remote from the Sun's warmth, mark the outer edge of the Solar System.

Our Galaxy

It may seem vast to us, but the Solar System is a tiny speck in the billions of stars in our Milky Way Galaxy. The Milky Way's nearest neighbors are two small galaxies called the Large and Small Magellanic Clouds. The Large Cloud (upper right) lies about 163,000 light-years away, while the Small Cloud is 195,000 light-years away.

Our Universe

The Milky Way (bottom left) is just one among many billions of galaxies in the universe. Despite its size—100,000 light-years across—our Galaxy is tiny compared to the universe as a whole. Light from even a close neighbor, the Andromeda Galaxy, takes about 2.5 million years to reach us.

Earth and Moon in Motion

The cycles of the sky give us our basic units of time—the day, month, and year. The tilt of our planet gives us the annual pulse of the seasons.

Earth's rotation on its axis provides us with the day. The turning of the Earth makes the Sun appear to travel across the sky from east to west. The interval from noon to noon, or midnight to midnight, defines the length of a solar day. It is this 24-hour-long day, one measured by the Sun, that we use to govern our daily lives.

ANOTHER KIND OF DAY

Astronomers use another type of day. Measure the time it takes a star to go from a point due south to that same point the next night, and you would find it takes 23 hours, 56 minutes and 4 seconds.

This is the length of a sidereal day, one measured by the stars, not the Sun. As a result of the shorter sidereal day, the stars rise nearly four minutes earlier each night. Over 30 days the difference adds up to two hours—a star that rises at 10 pm at the beginning of the month can be seen rising at 8 pm by month's end.

ZONES OF TIME

Until well into the 19th century, travelers could journey from town to town and find each place keeping a different clock time. The spread of railroads dictated a system of standard time. Introduced worldwide in 1884, this

Sunset
When we say "the Sun sets" or the "Sun rises" we are unconsciously invoking the ancient idea of a Sun that orbits a stationary Earth. It is, of course, Earth that is on the move, turning on its axis once every 24 hours, and making one complete orbit of the Sun in about 365 days.

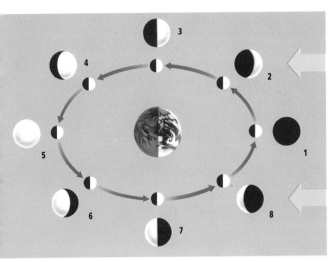

Waxing and Waning

At New Moon phase (1), when the Moon lies between us and the Sun, we cannot see the Moon at all. As the Moon swings away from the Sun to the east, we begin to see a waxing crescent phase (2). The nearside of the Moon becomes more and more daylit as the angle between Moon and Sun increases. A week after New, we see the Moon's disk half lit: First Quarter Moon (3). As the month continues, the Moon waxes to a gibbous, or Three-quarter Moon (4). Two weeks after New, the Moon lies opposite the Sun. The side of the Moon facing us now also faces the Sun and appears fully lit in our sky, as a Full Moon (5). In the next two weeks, the Moon enters its waning phases. The Moon decreases in phase from gibbous (6), to Last Quarter (7), then waning crescent (8) as it continues to orbit Earth and once again approaches the Sun in our sky. Just over 29 days after New Moon, the Moon once again comes between us and the Sun at New Moon phase and disappears from view.

system divides Earth into 24 time zones, each roughly 15 degrees wide, for a total of 360 degrees encircling the world. Within each zone the time is exactly the same but differs from the adjacent time zone by one hour. So when it is 12:32 pm in Boston it is 8:32 am in San Francisco, four time zones to the west.

The starting point of this worldwide system was originally known as Greenwich Mean Time (GMT), after the place in London where time was deemed to "begin" (see box, below). It is now known as Coordinated Universal Time (UTC). Times of astronomical events are often given in UTC. Calculating when that event might be visible from your location requires adding (if you live east of Greenwich) or subtracting (if you live west) the number of time zones, or hours, your home is from Greenwich; you would also have to take "daylight saving" or "summer time" (DST) into account.

GOING THROUGH A PHASE

While Earth's rotation gives us the day, the Moon's motion around Earth gives us the month. Measure the time it takes the Moon to return to the same spot in the stars and you get 27.32 days, a period known as the sidereal month. However, a more obvious cycle is the time it takes the Moon to go through a complete cycle of phases.

Just like Earth, half the Moon is always lit by the Sun and half is always dark. Contrary to popular

Only One View
The Moon takes the same time to rotate on its axis as it does to revolve around Earth. As a result, we only see one side of the Moon.

belief there is no perpetual "dark side of the Moon." Every place on the Moon experiences a cycle of lunar day and night. However, one side of the Moon does always face us—a braking force exerted by Earth's gravity has slowed the Moon's rotation so that one side of the Moon, its nearside, is locked to always face Earth.

As the Moon revolves around us, we see this lunar nearside lit up by different amounts, from none of it lit (New Moon), to half lit (Quarter Moon), to completely lit (Full Moon), creating the Moon's phases. A complete cycle of phases, from Full Moon to Full Moon for example, takes 29.53 days, the synodic month.

AN AWKWARD YEAR AND MONTH

Earth's motion around the Sun gives us our year. Earth takes 365.242199 days to revolve around the Sun, an awkward number that has plagued calendar makers since the beginning of measured time. Our modern-day Gregorian calendar uses 365 days as the closest practical length for the calendar year.

Now try fitting a whole number of synodic months (29.53 days) into a year. Impossible! The Gregorian calendar makes no attempt to do so—its month lengths of 30 and 31 days bear only a passing resemblance to the real length of the month, leading to oddities such as two Full

Moons in one month and 13 Full Moons in one year.

Other systems of counting time, such as the Hebrew and Islamic calendars, are governed by the phases of the Moon. In these calendars 12 lunar cycles define the year, giving a year 354 solar days long.

THE WOBBLING EARTH

In the Gregorian calendar, the year is defined as the time from March equinox to March equinox —the "tropical year" measured by the passage of the seasons. However, a year measured by how long it takes Earth to orbit the Sun with respect to the stars lasts 20 minutes and 24 seconds longer, the sidereal year.

The reason for the difference is that, like a spinning top, Earth wobbles through a slow precession motion that takes 25,800 years to complete. Today, Earth's rotation axis aims up to a point near the North Star, Polaris. Half a precession cycle from now, in AD 14,900, Earth's axis will point near the star Vega. Constellations

LEAPING YEARS

If every calendar year were just 365 days long, then after a few hundred years our calendar would get out of sync with the seasons. For the Northern Hemisphere, Christmas would come in the height of summer. In 45 BC, at the suggestion of Cleopatra and her court astronomer, Sosigenes, Julius Caesar introduced a calendar with a leap day, an extra day added every fourth year. The quadrennial leap day would work perfectly—if Earth took 365 $\frac{1}{4}$ days to go around the Sun. But the real year is 11 minutes and 14 seconds shy of that.

By the 16th century, the error in Caesar and Cleopatra's calendar had accumulated to 10 days. To correct this, in 1582 Pope Gregory XIII introduced refined leap year rules. Century years, ones ending in 00, would not be leap years, unless they were divisible by 400. So 1700, 1800, and 1900 were not leap years, but 1600 and 2000 were.

we now see in winter will be visible in summer, and current summer favorites will become wintertime constellations.

THE TILTED EARTH

The most popular misconception in astronomy is that the changing distance from Earth to the Sun causes the seasons. Earth's orbit is slightly elliptical, but that is not what causes the seasons. The tilt of the Earth does.

Midnight Sun
The tilted Earth causes summer's long days and winter's shorter ones. The effect is more extreme at the poles. Midsummer, with the pole tilted toward the Sun, is a time of continuous daylight; midwinter brings continuous night. A time-lapse photo taken in northern Finland shows the summer Sun's progress in the sky over two hours around midnight.

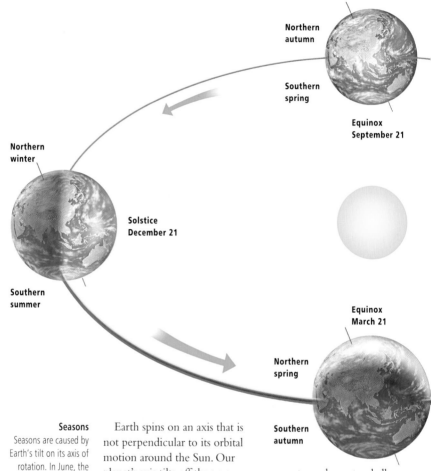

Northern
autumn

Southern
spring

Equinox
September 21

Northern
winter

Solstice
December 21

Southern
summer

Equinox
March 21

Northern
spring

Southern
autumn

Seasons

Seasons are caused by Earth's tilt on its axis of rotation. In June, the Southern Hemisphere is in winter because it is leaning away from the Sun, while the Northern Hemisphere is experiencing midsummer.

Six months later, the Northern Hemisphere is in midwinter, while the Southern Hemisphere is in midsummer. Earth's elliptical orbit has very little effect on the seasons.

Earth spins on an axis that is not perpendicular to its orbital motion around the Sun. Our planet's axis tilts off the perpendicular by 23.5 degrees and points to the same spot in the heavens no matter where our planet sits along its orbit.

On one side of our orbit, the Northern Hemisphere of Earth tilts toward the Sun. The Sun appears high in the sky, producing long, hot summer days. Half a year later, with Earth on the opposite side of the Sun, that same hemisphere is now tilted away from the Sun. Winter days are short and the Sun appears low in the sky. The Sun's energy enters

our atmosphere at a shallow angle, spreading the energy over a large surface area and diminishing its warming power.

TURNING POINTS OF THE YEAR

Four special dates punctuate the calendar. At one extreme, the summer solstice (June 21 in the Northern Hemisphere, December 21 in the Southern), the Sun climbs to its highest position in the daytime sky for the year. Daylight hours are at their maximum.

At the other extreme, winter solstice (December 21 in the

Seasonal Change
As the seasons progress, changes in day length and temperature set the vital rhythm for plants and animals, a rhythm most obviously seen in a deciduous tree's changing foliage.

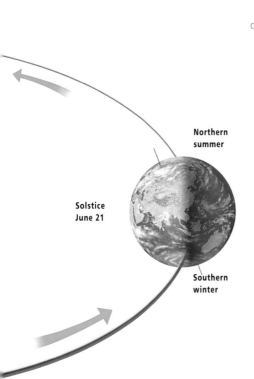

Northern summer

Solstice June 21

Southern winter

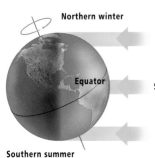

Northern winter

Equator

Southern summer

Sunlight

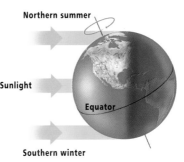

Northern summer

Equator

Southern winter

Spreading Sunlight
Sunlight strikes Earth at an angle in polar and mid-latitude zones, spreading out the heat and light across a wide area, and causing marked seasonal variations. The contrast beween the seasons becomes less marked closer to the Equator. There, the Sun's rays hit Earth almost squarely.

Northern Hemisphere, June 21 in the south), the Sun reaches its lowest noonday altitude. This is the shortest day of the year, with the minimum number of daylight hours.

On the two equinoxes (March 20 or 21 and September 22 or 23), neither hemisphere tilts toward or away from the Sun and all places on Earth experience 12 hours of daylight and 12 hours of night. The Sun rises due east and sets due west, while at noon the Sun lies halfway between its two extreme solstice altitudes.

Eclipses

At least four times a year, the Moon comes directly between us and the Sun, or we come between the Sun and the Moon, creating an eclipse visible somewhere on Earth.

Partial Solar Eclipse
In a partial solar eclipse, the Moon appears to take bite-size chunks from the Sun's disk. There is no "totality," and the solar corona remains invisible, unlike during a total eclipse.

Annular Eclipse
While the moon passes directly in front of the Sun in this time-lapse study of an annular eclipse, it is unable to cover the Sun completely, instead forming a distinctive ring shape.

In one of the great coincidences in nature, the Sun, 400 times the size of the Moon, also lies 400 times farther away than the Moon. As a result, both objects appear to be the same size in our sky, allowing the Moon to neatly cover the Sun in a total eclipse.

ECLIPSES OF THE SUN

Each month at New Moon phase, the Moon passes between us and the Sun. The tilt of the Moon's orbit usually takes it above or below the Sun's disk, so the Moon misses eclipsing the Sun. But at least twice a year the Moon's angled orbit intercepts the Sun's position just as the Moon reaches New phase. The three worlds fall into alignment with the Moon in the middle, and we get an eclipse of the Sun.

The eclipse might be partial, when the Moon covers only a

portion of the Sun. Or the eclipse could be annular: if the Moon is at the most distant point in its elliptical orbit, its disk will not be large enough to completely cover the Sun. We see the Sun reduced to a ring of light.

The most spectacular eclipse of all occurs when the Moon totally covers the bright disk of the Sun. The Moon's dark umbral shadow, often no more than 150 miles (240 km) wide, touches Earth and sweeps across the planet along a path several thousand miles long. People within this narrow track see a total eclipse, awe-inspiring but lasting only minutes.

A total solar eclipse occurs on average once every 19 months. For information about observing solar eclipses, see p. 236.

ECLIPSES OF THE MOON

At least twice a year the Sun, Earth, and Moon align with Earth in the middle. The Moon then passes through our planet's shadow, creating a lunar eclipse.

During a penumbral eclipse, the Moon passes through just the outer portion of Earth's shadow, and the Moon darkens so little that it is hard to detect that an eclipse is in progress. If the Moon grazes the dark central portion of Earth's shadow, the umbra, we see a partial lunar eclipse—the Full Moon looks like it has a dark bite taken from its bright disk.

About every 17 months (on average) the Full Moon passes completely into our planet's umbral shadow. The only sunlight that can reach the Moon during totality comes from red light filtering through Earth's atmosphere, and for up to two hours, the eclipsed Full Moon turns deep red. A total lunar eclipse can be seen from the entire side of Earth facing the Moon, allowing half the planet to enjoy the show.

For information about observing lunar eclipses, see p. 232.

Eclipse in Progress
With protective eyeglasses, in place, skywacthers track a solar eclipse.

Shadows and Light
A lunar eclipse (below left) occurs when the Moon passes through Earth's shadow. A solar eclipse (below right) happens when the Moon's shadow falls on Earth.

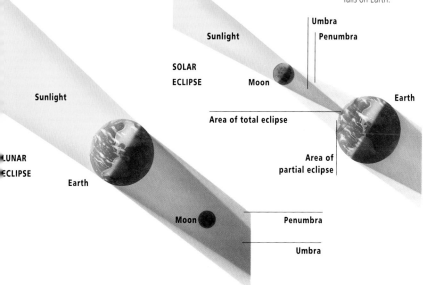

When the Sun Vanishes: a Total Solar Eclipse

A total eclipse of the Sun begins quietly, as the Moon first appears as a tiny bite on the side of the Sun's disk. Then the day grows dimmer as the Sun slowly disappears. When the eclipse is 99 percent total, the Sun appears like a brilliant star in a darkening sky. But not for long. The dark shadow of the Moon rises out of the east and races up the sky. As it reaches the Sun, the last flash of sunlight disappears behind the Moon.

Observers see a stunning sight: only the Sun's delicate outer atmosphere remains, surrounding the dark disk of the Moon, creating a black hole in a deep-blue sky.

The Sun Dies and Is Rebor
Time-lapse photography shows the progres of a total eclipse. At first contact (top left the Moon begins to encroach on the Su Over the next hour, during the initial parti phases, the Moon covers more of the Sun disk. Totality (central image) lasts for a fe unforgettable minutes, after which sun ligh begins to reappear. After an hour of parti phases, the Moon leaves the solar dis bringing the eclipse to an end

Black Disks and Diamond Rings

As totality approaches, the sky darkens to deep twilight. On the near-black disk of the Moon, a last burst of sunlight shines through a lunar valley creating a beautiful "diamond ring" effect (left). The corona, the normally invisible outer atmosphere of the Sun, suddenly appears at the moment of totality in the form of pearly white streamers (below).

The Celestial Sphere

On Earth we use longitude and latitude to determine location, and divide the continents into countries. Astronomers do something similar, mapping the sky into a gridwork pattern and into country-like regions, the constellations.

Celestial Sphere
The positions of stars and other heavenly bodies are described by a coordinate system gridded onto an imaginary celestial sphere. Declination and right ascension are used in much the same way as Earthly latitude and longitude, and the sky has an equator and two poles. The extra feature is the ecliptic—the apparent path of the Sun across the background of the celestial sphere. The red arrow indicates the sphere's apparent daily movement westward.

Though we know Earth does the moving, it can still be convenient to picture the sky as the ancients did, as a crystalline sphere turning above our heads with the sky's contents "pasted" to its interior surface. Astronomers have divided this sphere into a gridwork of coordinates.

THE SKY'S POLES AND EQUATOR

Imagine skewering the Earth, shish-kebab style, with a rod sticking from the South Pole of our planet, through its center, and out the North Pole. Then extend that rod in either direction far out into space. It hits our crystalline sphere at its north and south celestial poles. As Earth turns, the entire celestial sphere appears to rotate around these two poles.

Now, as if it were a giant elastic band, expand Earth's Equator line outward from Earth. It snaps onto our crystalline sphere to form the celestial equator, an imaginary line that divides the sky in half into its northern and southern hemispheres, just as Earth's Equator bisects our planet half-way between the poles.

MEASURING NORTH AND SOUTH—DECLINATION

Draw concentric lines parallel to the celestial equator, centered on the poles like a bull's-eye pattern. Like lines of latitude on Earth, these parallels of declination measure how far north or south of the Equator a sky location is. A star on the celestial equator has a declination of 0 degrees. A star at the north celestial pole has a declination of +90 degrees. A star halfway from the equator to the pole has a declination of +45 degrees. One degree (°) of declination contains 60 arcminutes ('), while each arcminute contains 60 arcseconds (").

MEASURING EAST AND WEST—RIGHT ASCENSION

Now draw lines running north and south from pole to pole. Like

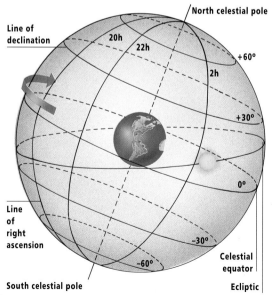

Line of declination

20h
22h
2h
+60°
+30°
0°
-30°
-60°

North celestial pole

Line of right ascension

South celestial pole

Celestial equator

Ecliptic

EPOCH
While all celestial bodies have right ascension and declination, these numbers slowly change because of a motion of the Earth called precession. As a result, catalogs and star charts specify a precise moment in time to define their coordinates. That moment is called the *epoch*.

ongitude on Earth, these lines of right ascension measure how far east or west a star is. The sky's equivalent of Earth's 0-degree Greenwich meridian is the point on the celestial equator where the Sun stands each year on the date of the March equinox. The line of right ascension that intersects the celestial equator at this point is defined as 0 hours right ascension (RA).

The sky is divided up into 24 hours (h) of right ascension, each hour containing 60 minutes (m) of time. The hours increase from west to east. For example, as Earth turns, a star near the celestial equator with an RA position of 4 hours will rise about four hours later than a star with an RA of 0 hours.

CELESTIAL COUNTRIES —THE CONSTELLATIONS
We can specify any location on Earth by giving its position in terms of latitude and longitude.

Outside the Sphere
This medieval-style French woodcut from the 19th century illustrates the old belief that Earth was surrounded by a vast star-studded dome. The man putting his head through the sphere symbolizes humanity's curiosity to see and understand what lies beyond our planet.

Orion the Hunter
Orion has been recognized as a distinctive group of stars for thousands of years. The Chaldeans of Mesopotamia knew it as Tammuz; the Syrians called it Al Jabbar, the Giant. To the ancient Egyptians it was Sahu, the soul of Osiris. It was the Greeks who gave the constellation its present name, commemorating a mythical giant and great hunter.

But it is easier to picture a city's location by thinking of it as belonging to a particular country. So it is with the sky. The bright star Betelgeuse has the celestial coordinates of +7° 24' declination and 5h 55.2m right ascension, but who can remember that? Instead, we say that Betelgeuse shines in the constellation of Orion the Hunter.

Learning to identify a dozen or so of the brightest constellations helps you find your way around the night sky. The stars cease to be a chaos of unrelated dots and begin to organize themselves into familiar and friendly patterns.

CONSTELLATIONS OLD AND NEW
In the 2nd century AD, the Greek astronomer Ptolemy recognized 48 ill-defined constellations. Today, we divide the sky into 88 constellations, with sharp borders drawn along zigzag boundaries officially agreed upon by astronomers in 1930. The boundaries may be relatively modern, but many constellations are as old as civilization itself. The ancient

merians, for example, depicted
lion in the sky where we still
e Leo and a bull where we
aw Taurus.

Other constellations are
ventions of 17th- and 18th-
ntury astronomers. Some were
eated to chart the southern
y, which was unknown to the
tronomers of Europe before the
e of exploration began in the
6th century. For example, Crux,
e Southern Cross, was first
efined as a constellation by
ndrea Corsali sailing to the
opics with a Portuguese
xpedition in 1515.

IN THE DOTS

he constellations are just
onvenient join-the-dot patterns
the sky. With rare exceptions,
e stars within a constellation
o not really belong together in
ace. For example, Betelgeuse—
e star forming Orion's east
oulder—lies 427 light-years
om Earth. Bellatrix, Orion's
est shoulder star, lies at half the

distance, 243 light-years away.
The stars of Orion's Belt shine
at us from a distance of about
800 to 900 light-years.

If we could travel hundreds of
light-years from Earth and view
Orion's stars from a different
place in the Milky Way Galaxy,
we would not see them forming
a stick-figure human as they do
from Earth. Aliens on other
planets would see very different
star patterns in their sky, although
their constellations might include
many of the same stars we see
here on Earth.

Indian Star Chart
Many old star charts are
more works of art than
practical guides to the
heavens. This beautiful
example was commissioned
by an Indian monarch in
1840. Though decorated
with Indian and Islamic
patterns, it features many
of the constellations we
use today.

Star Distances
Stars in a constellation look
as though they are all the
same distance from us.
This is an illusion—they are
almost always unrelated to
each other. The stars of
Orion, for example, range
in distance from Bellatrix
(243 light-years away) to
Mintaka and Alnilam (more
than 900 light-years away).

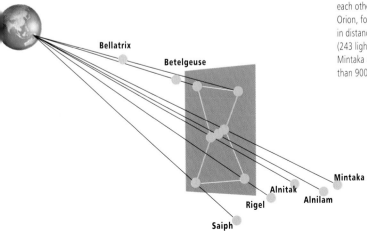

Bellatrix

Betelgeuse

Mintaka

Alnitak

Alnilam

Rigel

Saiph

The Motion of the Stars

At first glance, the stars seem to be fixed in place in the night sky, immovable. But a little observation over just a few hours will show that this is not so.

Star Paths by Latitude
The apparent motion of the stars depends on an observer's latitude.

Just as Earth's daily rotation carries the Sun from east to west across the sky during the day, each night the stars perform the same motion, circling the pole of the sky. Each year, as Earth revolves around the Sun, the stars and constellations also change with the seasons.

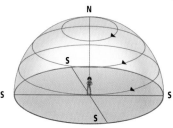

At the North Pole (90°N)

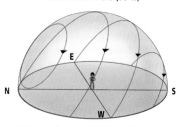

At the northern middle latitudes (40°N)

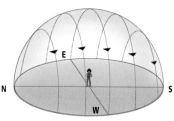

At the Equator (0°)

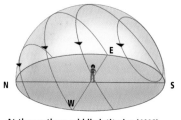

At the southern middle latitudes (40°S)

CHANGES THROUGH THE NIGHT
As the world turns, stars rise above the eastern horizon and set below the western horizon. On a December evening, for example, you might see Orion rising in the east. But by midnight Orion stands high in the sky. By dawn, he is setting in the west.

CHANGES WITH LATITUDE
As Earth spins, the sky appears to rotate around its two celestial poles. Just how the stars move in relation to your horizon depends on where on Earth you live (see diagram, left).

Stand on an ice floe at the North Pole and you would see the north celestial pole directly overhead. As Earth turns, the sky spins around this point. All the stars move in circles parallel to the horizon, never rising or setting. At the Equator, you will see the celestial equator directly overhead while the two poles of the celestial sphere lie on the horizon, due north and south. The sky rolls around these two horizon points, lifting stars straight up in the east and dropping them straight down to the horizon in the west.

Despite their nightly and seasonal cycles, the stars and constellations change little from year to year. To ancient astronomers 5,000 years ago, the Big Dipper and Orion looked the same as they do today. But the stars do slowly move as they, like the Sun, orbit the Milky Way Galaxy. Over tens of thousands of years, this *proper motion* of the stars will gradually distort our familiar constellations into unrecognizable shapes.

Between those two extremes, at mid-northern latitudes, the celestial sphere appears to be tilted over. Its north celestial pole lies halfway up the northern sky. Stars turn in circles centered on this stationary point due north. At the same latitude in the Southern Hemisphere, the stars still move east to west but turn around the south celestial pole, at a point partway up the sky due south.

CHANGES THROUGH THE YEAR

As Earth revolves around the Sun, the night side of the planet looks out toward a changing array of constellations—think of it as a giant carousel ride that sweeps us

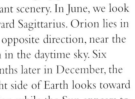

past a changing panorama of distant scenery. In June, we look toward Sagittarius. Orion lies in the opposite direction, near the Sun in the daytime sky. Six months later in December, the night side of Earth looks toward Orion while the Sun appears to be in Sagittarius.

Star Trails
Earth's rotation makes stars appear to move in the sky, a motion captured in this star-trail photo.

Star in a Hurry
Stars, like all objects in the universe, are on the move. This movement, or proper motion, can usually be discerned only over a period of thousands of years. But some stars are in a hurry. Kapteyn's Star in the constellation Pisces, for example, moves nearly 9 arcseconds per year. Its proper-motion shift is obvious in these two photos taken in 1975 (far left) and 1990 (left).

Getting Oriented: Finding North and South

The first step on any sky tour is to get oriented. To make sense of star charts you need to know which way is north and south, and for telescopes with equatorial mounts to track properly they must be aligned to the celestial pole. In the Northern Hemisphere look for Polaris, or the North Star. Though not the brightest star in the sky, Polaris is the brightest star closest to the north celestial pole and lies true north. In the Southern Hemisphere, with no bright "south star" to mark due south, use the Southern Cross to point to the sky's southern pole.

Finding the North Celestial Po[le]
In the Northern Hemisphere, locate the Big Dipp[er] in Ursa Major. Draw a line through the Pointe[r] Stars in the Dipper's bowl then extend tha[t] line away from the bowl's pouring lip. The lin[e] points to Polaris, the north celestial pol[e]

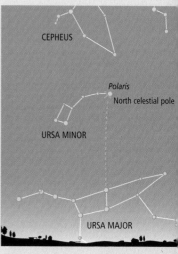

Finding the South Celestial Pole
In the Southern Hemisphere, locate Crux, the Southern Cross. Its long axis points to the south celestial pole, a blank area marked by dim Sigma Octantis. The pole lies halfway between Crux and the bright star Achernar and forms a right-angle triangle with brighter Canopus. A line perpendicular to a line joining Alpha and Beta Centauri also points to the pole. Crux, and Alpha and Beta Centauri feature in the photo on the facing page.

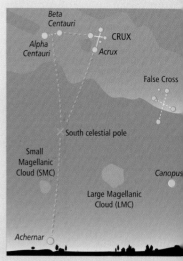

Measuring the Sky

Identifying a star or planet requires knowing how to estimate location in the sky and becoming familiar with the astronomer's scale of brightness. Once you do, almanac information such as "Jupiter, at magnitude –2.5, shines 5° north of the Moon" turns from jargon into useful information.

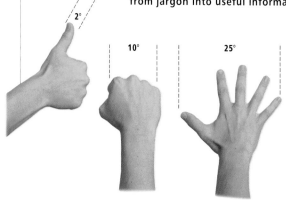

Hands Up

A simple way to measure distances in the sky is to use your hand, extended at arm's length. A thumb equals about 2 degrees, a fist 10 degrees, and a spread hand, 25 degrees.

How can you describe the size of the Big Dipper? Or its location? Astronomers use angular degrees. An angle of 90 degrees (abbreviated 90°) is the distance from horizon to zenith, the point overhead. An angle of 45 degrees takes you halfway up the sky. Hold your hand at arm's length and spread out your fingers and thumb. The distance from outstretched thumbtip to pinky fingertip is about 25 degrees. This is also the width of the Big Dipper and just less than the distance from the Dipper's Pointer Stars to the North Star.

Your fist at arm's length is about 10 degrees wide, your thumb about 2 degrees. You might think that the Moon is about as wide as your thumb at arm's length—but try it. You will discover that your thumb is four times wider. The Moon is only $^1/_2$ degree wide, or 30 arcminute (abbreviated 30'). The smallest angle you can discern ("resolve" is the astronomical term) with unaided eyes is about 1 arcminute, or 60 arcseconds (60"). The smallest object most back-yard telescopes can resolve is about 1 arcsecond (1") across.

THE SUN, THE CIRCLE, AND THE BABYLONIANS

Our method of dividing a circle into 360 degrees was invented by the Babylonians in about the 3rd millennium BC. They used a form of counting based on 60, not 100, and saw the circle as symbolic of the annual motion of the Sun, which takes about 360 days. It was natural for them to divide the circle into 360 degrees, with each degree being further subdivided into 60 seconds. We still use the Babylonian symbol for the Sun—a tiny circle—as shorthand for "degree." The same principle was applied to the concept of the day. The Babylonians divided it into two 12-hour periods, which were subdivided into 60-minute segments.

A QUESTION OF MAGNITUDE

Our system of measuring a star's brightness dates back to the 2nd century BC and the Greek astronomer Hipparchus. He ranked the brightest naked-eye stars as 1st magnitude and the dimmest as 6th. Today's system of apparent magnitude extends to brighter and fainter objects, but the rule remains the same: the fainter the object, the larger the number. Very bright objects have negative

magnitudes. The Sun blazes at magnitude −26.8. The brightest star in the night sky, Sirius, shines at magnitude −1.4. Vega is magnitude 0. The Big Dipper stars rank as 2nd magntude, while the faintest objects giant telescopes can detect glimmer at magnitude +30. A difference of five magnitudes equals a difference of a hundred times in brightness.

ABSOLUTELY BRILLIANT

Is the Sun really the brightest star in the sky? Place all the stars at the same distance from us then measure their brightnesses. What you get is a star's absolute magnitude. With the distance factor removed, you would see that stars still differ widely in brightness.

Astronomers define absolute magnitude as the apparent magnitude a star would have if it were 10 parsecs or 32.6 light-years from us (see box, p. 42). The Sun would shine with an absolute magnitude of +4.8, nearly 10 magnitudes, or 10,000 times, dimmer than the supergiant star, Betelgeuse, with its absolute magnitude of −5.0.

Apparent vs Absolute
Regulus, the blue-white star at lower right in Leo, appears nearly the same brightness as orange Algieba diagonally above it. Yet Regulus is twice as far away from Earth as Algieba. They look the same only because Regulus has a much greater absolute magnitude.

Magnitude Scale
The apparent magnitude scale describes how objects *appear* in our sky; it does not describe the objects' true brightness.

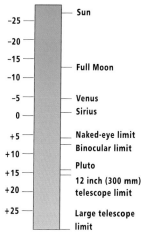

Magnitude	
−25	Sun
−20	
−15	
−10	Full Moon
−5	Venus
0	Sirius
+5	Naked-eye limit
+10	Binocular limit
+15	Pluto
+20	12 inch (300 mm) telescope limit
+25	Large telescope limit

Names in the Sky

Examine a star chart and you will discover celestial places labeled with strange-sounding names (Betelgeuse, Aldebaran, Rigel) and technical sounding numbers (SAO 113271, M 42, NGC 4565). Where do the names come from? And what do the code numbers mean?

Flamsteed Numbers
Below right: John Flamsteed introduced a new system of numbering stars in his catalog *Historia Coelestis Britannica*, published in 1725.

What's in a Name?
Below left: Many stars and deep-sky objects have more than one name. The Orion Nebula is also known as the Great Nebula, M 42, and NGC 1976.

Of the 6,000 stars visible to the unaided eye, about 250 have proper names. Some, like Aldebaran, are Arabic in origin—it means "the Follower," from this star's position in the sky following the Pleiades star cluster. Other star names, such as Arcturus, are Greek and link to the ancient myths associated with the constellations. Arcturus means "the Bear Watcher or Guardian," referring to this star's proximity to the constellations Ursa Major and Ursa Minor, the Great and Little Bears.

THE NUMBERING OF STARS
As in a big city with only names for streets, finding your way around the sky would be difficult if we used only names. In the early 1600s, the German astronomer Johann Bayer applied Greek letters to stars. He labeled the brightest star in a constellation alpha (α), the second brightest beta (β), and so on. Arcturus, the brightest star in Boötes, for example, is also known as Alpha Bootis. (By convention in such designations, the constellation name takes the Latin genitive case, so Boötes becomes Bootis, meaning "Alpha of Boötes.")

With only 24 letters in the Greek alphabet, we soon run out of Bayer letters. In the 1700s, the English astronomer John Flamsteed numbered stars within a constellation from west to east. So Arcturus is also known in

Flamsteed's system as 16 Bootis. Today, modern catalogs such as the Smithsonian Astrophysical Observatory (SAO), Hipparcos (HIP), Tycho (TYC), and Hubble Guide Star Catalog (GSC) have numbered millions of stars.

A CATALOG OF NON-COMETS

In the late 1700s, the French astronomer Charles Messier created a catalog of more than a hundred objects that still form the heart of any backyard astronomer's observing list. Messier charted some of the sky's best star clusters, nebulas, and galaxies—the Andromeda Galaxy, for example, is known as M 31.

Messier wanted to avoid confusing these objects with comets, the real target of his nightly star hunts. Today, we remember Messier not for the comets he found but for his catalog of the deep-sky objects he considered nuisances.

NGC WHAT?

By the late 1800s, with the growing size of telescopes, astronomers had explored far beyond Messier's simple list, charting thousands of deep-sky objects. In 1888, the British astronomer J. L. E. Dreyer collated an assortment of lists into one master catalog, the New General Catalogue or NGC. Along with two supplementary Index Catalogues (IC), Dreyer's lists contain more than 13,000 clusters, nebulas, and galaxies suitable for backyard telescopes. For example, M 31 is also known as NGC 224.

Constellation Rethink
Not all constellations pass the test of time. Carina, Vela, and Puppis once formed a constellation called Argo Navis, the Ship Argo—the vessel of Jason and the Argonauts—shown here in a 1600s star atlas.

Messier Objects
Charles Messier charted 103 fuzzy-looking objects he encountered on his search for comets in a catalog published in 1781.

Resources
Guide

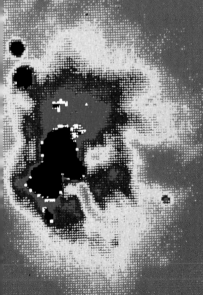

Resources Guide

WEBSITES

The web is the first place to turn to stay current with astronomy and space exploration. NASA and other countries' space agencies now release most of their information through the web, as do major world observatories. (For brevity, "http://" has been omitted in the list below.)

AstroWeb
www.cv.nrao.edu/fits/www/ astronomy.html This is a classified index to astronomical websites. All the links are tested daily. This is best place to look for individual observatories' websites.

NASA Watch
www.nasawatch.com Not a NASA-sponsored website, but a generally reliable place to check for news about the space program that has not yet broken officially.

NASA Today
www.nasa.gov/today/index.html A quick place to look for what's new.

Jet Propulsion Laboratory
www.jpl.nasa.gov JPL runs many of NASA's unmanned missions, including those to Saturn (Cassini), Jupiter (Galileo), and Mars. This is the place to check the latest findings.

European Space Agency
sci.esa.int Designs and flies many missions that often dovetail with NASA's.

European Southern Observatory
www.eso.org Has four 8.4 m telescopes at high altitude in Chile—their website has outstanding images, taken with their new large telescopes.

Shuttle flights and the International Space Station
spaceflight.nasa.gov

Unmanned space missions
spacescience.nasa.gov/missions/ index.htm Links to all missions, past and present, plus future ones that have started the design process.

Hubble Space Telescope
oposite.stsci.edu This public information site has links to all images released by HST.

X-ray astronomy
chandra.harvard.edu

Planetary information
www.seds.org/nineplanets/ nineplanets/nineplanets.html The Nine Planets site has descriptive guides to the Solar System bodies.
nssdc.gsfc.nasa.gov/planetary/ planetary_home.html This site has descriptions and provides links to spacecraft mission descriptions and data.

Auroras and other "space weather"
spaceweather.com
www.sel.noaa.gov

Currently known extrasolar planets
www.obspm.fr/encycl/encycl.html Updated as often as new results appear, this is the best place to check on the state of this rapidly changing field.

Deep-sky objects
www.seds.org/messier A semi-encyclopedic guide to the brightest and best deep-sky galaxies, star clusters, nebulas. It has links to lots of images.

NASA's database on galaxies
nedwww.ipac.caltech.edu Aimed at professional astronomers, this site is still easy to use. Start with the help page, though.

A directory of astronomy clubs, planetariums, and observatories
www.skypub.com/resources/ directory/directory.html Worldwide in scope, this is organized by country and state, with links to websites (where they exist). Also has a link to star parties.

The Search for Extraterrestrial Intelligence
www.seds.org/~rme/seti.html
www.seti.org

PRINTED RESOURCES

While some of the following are available in new bookstores, check out used bookstores as well. And become familiar with the astronomy and space science collections in your local libraries, both public and university.

Armchair Astronomy
The Cambridge Atlas of Astronomy, Jean Audouze, Guy Israël (Cambridge University Press, 1994). A massive and thoroughly illustrated encyclopedia of all areas of astronomy.
A Concise Dictionary of Astronomy, Jacqueline Mitton (Oxford University Press, 1991).
Galaxies, Tim Ferris (Sierra Club, 1980). Whether in the original edition or the smaller-size

reprints, this is a stunningly beautiful book full of photos of all kinds of galaxies

Magnificent Universe, Ken Croswell (Simon & Schuster, 1999). A beautifully illustrated tour of the cosmos.

The Universe and Beyond, Terence Dickinson (Firefly Books, 1999). A marvelous work that combines photos and space art to portray the universe in all its grandeur.

Star History and Mythology

History of Astronomy, Antonie Pannekoek (Dover, 1989). Despite its age, this is the best general history of astronomy from antiquity up to the 1930s.

Stars of the First People, Dorcas S. Miller (Pruett Publishing, 1997). An outstandingly detailed and well-organized tour of Native American star myths and constellation figures. Thoroughly researched (with citations to ethnography literature) and very readable. Organized by region, and by tribe within regions.

Star Tales, Ian Ridpath (Universe Books, 1988). A highly readable guide to the Greek and Roman myths surrounding the constellations we still use.

Old Astronomy Maps

Celestial Charts, Carole Stott (Smithmark, 1995). A beautifully illustrated tour of publications that few of us will ever own.

The Mapping of the Heavens, Peter Whitfield (Pomegranate, 1995). An equally wonderful guide whose illustrations don't overlap much with Stott's.

The Space Race

The Case For Mars, Robert Zubrin (Free Press, 1996). Why we should explore and settle the

Red Planet, written by a strong partisan.

Entering Space, Robert Zubrin (Tarcher/Putnam, 1999). The why and the how of establishing a "spacefaring civilization." Powerfully visionary.

A Man on the Moon, Andrew Chaikin (Viking, 1994). An excitingly personal look at the Apollo Moon-landing program taken from extensive interviews with the astronauts.

This New Ocean, William E. Burrows (Random House, 1998). A clear-eyed examination of the coalition of interests—military, political, and visionary—that created the Space Age.

Observing Guides

These are listed from simplest to more detailed.

A Skywatcher's Year, Jeff Kanipe (Cambridge University Press, 1999). Taking the sky week by week, 52 easy and comfortable essays provide a clear introduction to the night sky through all the seasons.

Night Sky, Discovery Channel (Discovery Books, 1999). Handy, pocket-size guide to astronomy and observing.

Skywatching, David H. Levy (The Nature Company, 1994). A beautifully illustrated first step into astronomy.

Advanced Skywatching, Robert Burnham, Alan Dyer, Robert A. Garfinkle, Martin George, Jeff Kanipe, David H. Levy (The Nature Company, 1997). This takes the reader deeper into backyard stargazing; highly readable.

The Backyard Astronomer's Guide, Terence Dickinson, Alan Dyer (Camden House, 1991). A guide to all aspects of amateur astronomy, with

especially good advice on shopping for equipment.

Binocular Astronomy, Craig Crossen (Willmann-Bell, 1992). Binoculars often serve as a first telescope, and can show you a lot when you know what you're looking at. Crossen provides an informative guide.

Star-Hopping, Robert A. Garfinkle (Cambridge University Press, 1997). This provides year-round tours for exploring all areas of the sky visible from middle northern latitudes.

Hartung's Astronomical Objects for Southern Telescopes, David Malin, David J. Frew (Cambridge University Press, 1995). Brings up to date a classic observing guide for Southern Hemisphere observers.

Observing Handbook and Catalogue of Deep-Sky Objects, Christian B. Luginbuhl, Brian A. Skiff (Cambridge University Press, 1990). Just what the title says, written by two highly experienced observers.

The Night Sky Planisphere, David Chandler (1992). A planisphere is a flat, circular plastic skymap that shows you the stars visible at a given latitude for any date and time in the year. Many different makes exist. Available for several latitudes, north and south, through *Sky & Telescope* magazine (see below).

Star Atlases and Catalogs

The Cambridge Star Atlas, Wil Tirion (Cambridge University Press, 1996). Best suited for binoculars and small telescopes, this has 20 maps covering the sky, with lists of deep-sky objects on the facing pages.

The Guide Star Catalog (NASA, 1991). These two CD-ROMS

were developed for aiming the Hubble Space Telescope, and contain data on about 19 million stars and objects all over the sky.

NGC 2000.0, Roger W. Sinnott, (Cambridge University Press/ Sky Publishing, 1988). This catalog of non-stellar objects updates and replaces Dreyer's long-outdated *New General Catalogue* and *Index Catalogues.*

Sky Atlas 2000.0, Wil Tirion (Cambridge University Press, 1998). The standard star atlas used by most observers; it shows the entire sky in 26 large fold-out charts.

Sky Catalogue 2000.0, Alan Hirshfeld, Roger W. Sinnott (Cambridge University Press, 1991 & 1985). A two-volume work listing basic data such as positions, brightnesses, sizes, distances, etc. for stars, galaxies, star clusters, and nebulas. Designed to accompany *Sky Atlas 2000.0.*

Uranometria 2000.0, Wil Tirion, Barry Rappaport, George Lovi (Willmann-Bell, 1987 & 1988). In two volumes, this goes fainter than *Sky Atlas 2000.0.* A matching *Deep-Sky Field Guide to Uranometria 2000.0* (Willmann-Bell, 1992) gives basic information for objects on the *Uranometria* charts.

Annual Almanacs

Astronomical Almanac (U.S. Government Printing Office, annual). The standard tables of planet positions and data for each year, available in many libraries.

Astronomical Calendar, Guy Ottewell (Astronomical Workshop, annual). A highly graphic way to explore the astronomical events of each year. Ottewell's illustrations

will change how you think of the sky.

Observer's Handbook, Roy L. Bishop (Royal Astronomical Society of Canada, annual). This is a handy and convenient annual compendium of sky events and data.

Telescope Building

Most people today buy a ready-made telescope instead of building their own. For anyone wanting to make a customized telescope, or who simply wants to recapture the feel of yesteryear, the following publications provide clear directions. For reviews of telescopes and equipment, check issues of *Sky & Telescope* and *Astronomy.*

Amateur Radio Astronomer's Handbook, John Potter Shields (Crown Publishers, 1986). Provides how-to instructions for building and hooking up radio receivers, chart recorders, antennas, and other equipment.

Build Your Own Telescope, Richard Berry (Willmann-Bell, 1994). Provides detailed directions for five telescopes that will give fine performance and which can be built using simple tools.

The Dobsonian Telescope, David Kriege, Richard Berry (Willmann-Bell, 1997). Tells how to build a 14 inch (350 mm) reflector on a simple, low-cost mounting.

Astro-Imaging

As with ordinary picture-taking, astronomical photography divides between conventional film and electronic imaging. These books will get you started down whichever path appeals most.

Astrophotography for the Amateur, Michael A. Covington

(Cambridge University Press, 1999). An excellent guide for beginning astrophotographers.

The CCD Camera Cookbook, Richard Berry, Valko Kanto, John Munger (Willmann-Bell, 1994). Tells how to build your own simple but fully functional CCD (electronic) camera.

Choosing and Using a CCD Camera, Richard Berry (Willmann-Bell, 1992). Explains what CCD cameras are and how they work, for the beginner.

The Handbook of Astronomical Image Processing, Richard Berry, James Burnell, (Willmann-Bell, 2000). Electronic imaging is astro-photography's future. This book (which comes with a CD-ROM) provides both software and sample images to work with on a Windows computer.

Splendours of the Universe, Terence Dickinson, Jack Newton (Firefly, 1997). A very basic, but quite thorough introduction to skyshooting—beautifully illustrated.

Exploring the Solar System

In the following, general Solar System works come first, then ones that concentrate on each planet outward from the Sun.

Moons and Planets, William K. Hartmann (Wadsworth, 1998). A very readable and well illustrated textbook on how planets and moons formed and evolved.

The New Solar System, J. Kelly Beatty, Carolyn Collins Petersen, Andrew Chaikin (Sky Publishing, 1999). Chapters on nearly all the individual planets and bodies in the Solar System, written for the lay reader.

Introduction to Observing and Photographing the Solar System, Thomas A. Dobbins,

Donald C. Parker, Charles F. Capen (Willmann-Bell, 1988). A practical handbook on Solar System observing with a backyard telescope.

Observing the Sun, Peter O. Taylor (Cambridge University Press, 1991). Explains how to view the Sun safely with any telescope, and describes in straightforward terms how the Sun works.

Mercury: the Elusive Planet, Robert G. Strom (Smithsonian Institution Press, 1987). A readable description of a planet whose surface looks lunar but whose interior is more like Earth's.

Venus Revealed, David Harry Grinspoon (Addison-Wesley, 1997). Excellent and highly readable account by a Venus scientist.

The Once and Future Moon, Paul Spudis (Smithsonian Institution Press, 1996). Describes the Moon's geological history; written by a lunar researcher.

Observing the Moon, Gerald North (Cambridge University Press, 2000). A backyard observer's guide that examines 48 selected features in detail.

Atlas of the Moon, Antonín Rükl (Hamlyn, 1994). A must for every Moon-watcher; beautiful charts of the lunar nearside.

Uncovering the Secrets of the Red Planet Mars, Paul Raeburn (National Geographic Society, 1998). A beautifully illustrated guide to Mars as seen by the Mars Pathfinder lander and Sojourner rover.

The Planet Mars, William Sheehan (University of Arizona Press, 1996). A highly readable history of Martian studies.

Introduction to Asteroids, Clifford Cunningham (Willmann-Bell, 1988). A detailed guide to

asteroids from a physical and observational point of view.

The Giant Planet Jupiter, John H. Rogers (Cambridge University Press, 1995). A thorough and splendidly illustrated guide to Jupiter, with an emphasis on observations from Earth.

The Planet Saturn, A. F. O'D. Alexander (Dover, 1980). A historical survey of pre-spacecraft findings—just the thing for a cloudy night.

Voyages to Saturn, David Morrison (NASA, 1982). Blends you-are-there history of the Voyager flybys with the initial science results; lots of photos.

The Planet Uranus, A. F. O'D. Alexander (Faber and Faber, 1965). A guide to the Earth-based observational history of the planet through the 1950s.

Uranus, Ellis Miner (Ellis Horwood, 1997). Brings the story up through the Voyager 2 encounter in 1986, with all its scientific findings presented clearly.

Atlas of Neptune, Garry E. Hunt, Patrick Moore (Cambridge University Press, 1994). A well-illustrated guide to the planet and its moons which includes the findings from the Voyager flyby in 1989.

Out of the Darkness; the Planet Pluto, Clyde W. Tombaugh, Patrick Moore (Stackpole Books, 1980). Clyde Tombaugh's personal story

of how he discovered this small, distant world.

Great Comets, Robert Burnham (Cambridge University Press, 2000). Focuses on the brightest and rarest comets seen, including Comet Hale-Bopp and Hyakutake.

Observing Comets, Asteroids, Meteors, and the Zodiacal Light, Stephen J. Edberg, David H. Levy (Cambridge University Press, 1994). A practical how-to book aimed at all levels of skygazers.

Thunderstones and Shooting Stars, Robert T. Dodd (Harvard University Press, 1986). Well-written short guide to meteorites and what they tell about the Solar System's history.

Other Planetary Systems

Planet Quest, Ken Croswell (Free Press, 1997). An easily read guide to how astronomers began finding planets orbiting other stars, and what these new-found solar systems have to tell us about our own.

Stars & Galaxies

Alchemy of the Heavens, by Ken Croswell (Anchor Books, 1996). A well-written account of the galaxy we live in.

Observing Variable Stars, David H. Levy (Cambridge University Press, 1989). How amateurs with backyard telescopes can follow variables.

Planetary Nebulae, Steven J. Hynes (Willmann-Bell, 1991). A practical guide for sky-watchers, with lists, sketches, and finder charts.

The Supernova Search Charts and Handbook, Gregg Thompson, James Bryan, Jr. (Cambridge University Press, 1989). Charts for more than 300 galaxies help observers hunt for supernovas in them.

Cosmology

Just Six Numbers, Martin Rees (Basic Books, 2000). The six numbers of this book's title are physical parameters that have shaped the universe we live in; had they been only slightly different, we wouldn't be here.

Lonely Hearts of the Cosmos, Dennis Overbye (HarperCollins, 1991). A wonderfully personal look at cosmology and its practitioners. Highly enjoyable.

The Runaway Universe, Donald Goldsmith (Perseus Books, 2000). A highly readable look at a startling recent discovery by cosmologists—that the expansion of the universe is accelerating.

The Whole Shebang, Timothy Ferris (Simon & Schuster, 1997). Subtitled "a state of the universe report," this outlines the current theories for the general reader.

Is Life Unique to Earth?

Are We Alone? Paul Davies (Basic Books, 1995). This book for the general reader is about the philosophical implications for the discovery of extra-terrestrial life.

Here Be Dragons, David Koerner, Simon LeVay (Oxford University Press, 2000). Written for the general reader, this covers how life emerged on Earth and what are its chances for existing elsewhere in the universe.

Rare Earth, Peter Ward, Donald Brownlee (Copernicus Books, 2000). Is Earth, with its diversity of lifeforms, an anomaly or not? This book aimed at the general reader concludes, unlike most books on the subject, that Earth-like planets and life are probably quite uncommon.

The Search for Life on Other Planets, Bruce Jakosky (Cambridge University Press, 1998). Similar to the above, this provides a more detailed look at the science while remaining highly readable.

SOFTWARE

These days, most backyard astronomers have at least one program on their computers to show what is visible in the sky from any place at any date and time. Many programs exist in all price categories, but these two are the standards by which all others are judged.

Starry Night, Space.com.
An outstanding program with an exceptionally easy-to-use interface and a realistic sky. It is available in simplified or detailed versions. Mac and Windows.
tel: +1 800-221-8180
www.starrynight.com

TheSky, Software Bisque.
A powerful program for Windows or Mac computers. It features plug-ins to allow telescope control from the keyboard.
tel: +1 303-278-4478
www.bisque.com

MAGAZINES

Sky & Telescope is the oldest astronomy magazine for backyard skygazers, while many beginners find *Astronomy*'s features more accessible. Both monthly periodicals are essential. *Mercury* is a popular-level bimonthly put out by the Astronomical Society of the Pacific. *SkyNews* is a popular-level bimonthly for Canadian observers. *The Planetary Report* is a bimonthly that covers developments in Solar System exploration for lay readers.

Important news in astronomy and space exploration appears in two weeklies aimed at scientists, *Science* and *Nature*. Both usually have a less-technical commentary accompanying the scientific papers. *Science News* and *New Scientist* are weeklies aimed at the general reader, and almost every issue has astronomy and space-related stories. The same goes for the monthlies *Discover* and *Scientific American*. For children, check out *Odyssey*, and *Scientific American Explorations*. A trip to a university library will uncover many publications for professional astronomers and planetary scientists.

Astronomy, P.O. Box 92788, Waukesha, WI 53187, USA.
tel: +1 262-796-8776
www.astronomy.com

Discover, 111 Fifth Ave., New York, NY 10011, USA.
tel: +1 800-829-9132
www.discover.com

Mercury, 390 Ashton Ave., San Francisco, CA 94112, USA.
tel: +1 415-337-1100
www.aspsky.org

Nature, 4 Crinan St., London N1 9XW, England.
tel: +44 20-7843-4626
www.nature.com

New Scientist, 151 Wardour St., London W1V 4BN, England.
tel: +44 20-8652-3500
www.newscientist.com

Odyssey, 30 Grove St., Suite C, Peterborough, NH 03458, USA.
tel: +1 603-924-7209
www.odysseymagazine.com
The Planetary Report, 65 N. Catalina Ave., Pasadena, CA 91106, USA.
tel: +1 626-793-5100
www.planetary.org
Science, 1200 New York Ave., N.W., Washington, DC 20005, USA.
tel: +1 202-326-6501
www.sciencemag.org
Science News, 1719 N St., N.W., Washington, DC 20036, USA.
tel: +1 202-785-2255
www.sciencenews.org
Scientific American, 415 Madison Ave., New York, NY, 1001, USA.
tel: +1 800-333-1199
www.sciam.org
Scientific American Explorations, 179 South St., 6th Floor, Boston, MA 02111, USA.
tel: +1 617-338-8231
www.explorations.org
Sky & Telescope, P.O. Box 9111, Belmont, MA 02178-9111, USA.
tel: +1 617-864-7360
www.skypub.com
SkyNews, Box 10, Yarker, Ontario, Canada K0K 3N0
www.skynewsmagazine.com

ASTRONOMY ORGANIZATIONS

To find an astronomy club in your area, ask at the public library or a local college or university. Also, check the annual special issues published by *Astronomy* and *Sky & Telescope*. The magazine websites also have current information, organized by state or country.

American Association of Variable Star Observers (AAVSO), 25 Birch St., Cambridge, MA 02138, USA. Members (mostly amateurs) help professional astronomers by using backyard telescopes to monitor thousands of stars that change in brightness.
tel: +1 617-354-0484
www.aavso.org

American Astronomical Society (AAS), 2000 Florida Ave., NW, Suite 400, Washington, DC 20009, USA. Although it is for professionals, the AAS organizes a lot of public outreach and education efforts. The website has excellent info on how to become an astronomer.
tel: +1 202-328-2010
www.aas.org

Association of Lunar and Planetary Observers (ALPO), Harry Jamieson, P.O. Box 171302, Memphis, TN 38187, USA. An organization of backyard telescope users primarily interested in the Solar System.
www.lpl.arizona.edu/alpo

Astronomical Society of the Pacific (ASP), 390 Ashton Ave., San Francisco, CA 94112, USA. An international organization of amateur and professional astronomers.
tel: +1 415-337-1100
www.aspsky.org

European Space Agency (ESA), 8–10 rue Mario-Nikis, Paris Cedex 15, France
tel: +33 1-5369-7155
www.esa.int

International Dark-Sky Association, 3545 N. Stewart St., Tucson, AZ 85716, USA. An organization dedicated to helping astronomers (professional and amateur) preserve dark skies for observing.
www.darksky.org

International Meteor Association (IMO), Robert Lunsford, 161 Vance St., Chula Vista, CA 91910, USA. Coordinates meteor observing programs and runs an e-mail list server to keep members and interested people up-to-date.
www.imo.net

National Aeronautics and Space Administration (NASA), 300 E St., S.W., Washington, DC 20546, USA.
tel: +1 202-358-1588
www.nasa.gov

The Planetary Society 65 N. Catalina Ave., Pasadena, CA 91106, USA. This international organization promotes Solar System exploration using both unmanned spacecraft and human expeditions. There are chapters across the U.S. and in other countries.
tel: +1 626-793-5100
www.planetary.org

Royal Astronomical Society of Canada (RASC), 136 Dupont St., Toronto, Ontario, Canada M5R 1V2. Canada's national astronomy organization covers professional and amateur astronomers; it has local "centres" (chapters) in many cities.
tel: +1 416-9247973
www.rasc.ca

STAR PARTIES

Star parties are get-togethers for amateur astronomers, usually held over a weekend around New Moon in a location free of intruding lights. They are lots of fun and you don't even have to bring a telescope. Because contact people often change year to year, and dates shift to avoid a bright Moon, it's always best to check online with magazines such as *Astronomy* (www.astronomy.com) or *Sky & Telescope* (www.skypub.com) for the latest information.

Glossary

Active galaxy A galaxy with a central black hole that is emitting lots of radiation.

Aperture The diameter of a telescope's main light-collecting optics, whether a lens or a mirror. Also, the diameter of a binocular lens.

Arcminute A unit of angular measure equal to ¹⁄₆₀ of a degree; the Moon and Sun are about 30 arcminutes across.

Arcsecond A unit of angular measure equal to ¹⁄₆₀ of an arcminute; Jupiter averages some 44 arcseconds across.

Astronomical unit (AU) The average distance between Earth and the Sun, about 93 million miles (150 million km).

Atmosphere The layer of gases attached to a planet or moon by gravity.

Axis The imaginary line through the center of a planet, star, or galaxy around which it rotates; also, a shaft around which a telescope mounting pivots.

Big Bang The explosion of a small, very hot lump of matter about 15 billion years ago that marked the birth of the universe, according to the current theory of the universe's origin.

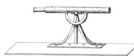

Binary star (double star) Two stars linked by mutual gravity and revolving around a common center of mass, in contrast to a chance alignment of two stars as seen from Earth.

Black hole A massive, compact object so dense that no light

or other radiation can escape from inside it.

Celestial equator The imaginary line encircling the sky midway between the celestial poles.

Celestial poles The imaginary points on the sky where Earth's rotation axis, extended infinitely, would touch the imaginary celestial sphere.

Celestial sphere The imaginary sphere enveloping Earth upon which the stars, galaxies, and other celestial objects all appear to lie.

Comet A small body composed of ices and dust which orbits the Sun on an elongated path.

Constellation One of the 88 official patterns of stars that divide the sky into sections.

Declination The angular distance of a celestial object north or south of the celestial equator. It corresponds to latitude on Earth.

Degree A unit of angular measure equal to ¹⁄₃₆₀ of a circle.

Electromagnetic spectrum The full range of radiation produced by nature; runs from high-energy gamma rays to very long radio waves.

False-color An enhancement technique used in image processing to make subtle differences stand out.

Focal length The distance between the main lens or mirror of a telescope and the point where the light from it comes to a focus.

Galaxy A huge gathering of stars, gas, and dust, bound by gravity and having a mass ranging from 100,000 to 10 trillion times that of the Sun.

Gamma rays Radiation with wavelength shorter than X rays.

Gas-giant planet A planet whose composition is dominated by hydrogen: Jupiter, Saturn, Uranus, and Neptune.

Globular star cluster A spherical cluster that may contain up to a million stars.

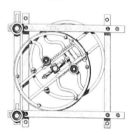

Gravitational lens A galaxy or other massive object standing between Earth and a more distant object. Its gravity bends the light from the distant object and creates distorted or multiple images of it.

Impact crater The round, bowl-like scar left on the surface of a moon or planet when it is struck by a meteorite.

Infrared (IR) Radiation with wavelengths just longer than visible light.

Light-year The distance that light travels in one year, about 6 trillion miles (9.5 trillion km).

M objects Star clusters, nebulas, and galaxies in the Messier list, compiled by 18th-century comet-hunter Charles Messier to create a roster of objects that resembled comets.

Magnetic field The region of space in which a celestial object exerts a magnetic force.

Magnitude A unit of brightness for celestial objects. Apparent magnitude describes how bright a star looks from Earth, while absolute magnitude is its brightness if placed at a distance of 32.6 light-years.

Meteor The bright, transient streak of light produced by a meteoroid, a piece of space debris, burning up as it enters Earth's atmosphere at high speed.

Meteorite Any piece of inter-planetary debris that reaches Earth's surface intact.

Microwave Radiation with wavelengths measured in millimeters.

Nebula A cloud of gas or dust in space; may be either dark or luminous.

NGC objects Galaxies, star clusters, and nebulas listed in the *New General Catalogue* of J. L. E. Dreyer, published in the late 19th century.

Nova A white dwarf star in a binary system that brightens suddenly by several magnitudes as gas pulled away from its companion star explodes in a thermonuclear reaction.

Nucleus The central core of a galaxy or comet.

Open star cluster A group of some few hundred stars bound by gravity and moving through space together.

Optical Radiation with wavelengths detectable by the human eye.

Orbit The path of an object as it moves through space under the control of another's gravity.

Planetary nebula A shell of gas puffed off by a star late in its life. Their often round appearance led to the name.

Pulsar An old, rapidly spinning star that flashes periodic bursts of radio (and occasionally optical) energy.

Quasar A quasi-stellar object or radio source whose spectrum shows a high velocity of recession from Earth; quasars are thought to be the active nuclei of very distant galaxies.

Radiant The point on the sky from where a shower of meteors appears to come.

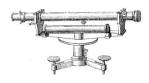

Radiation The means by which energy travels through space; it has characteristics of both waves and particles.

Radio Radiation of centimeter or longer wavelength.

Radio astronomy The study of celestial bodies by means of the radio waves that they emit and absorb.

Red giant A large, cool, red star in a late stage of its life.

Redshift An apparent shift of spectral lines in the light of an object toward the red end of the spectrum, caused by relative motion between the object and Earth.

Right ascension (RA) The celestial coordinate analogous to longitude on Earth.

Satellite Either a moon or a spacecraft in orbit around a planet.

Seeing A measure of the steadiness of the atmosphere. Good seeing is essential to using high magnifications.

Sunspot A dark, highly magnetized region on the Sun's surface, cooler than the surrounding area.

Supernova The massive explosion of a star in which it blows off its outer atmosphere and briefly equals a galaxy in brightness.

Tectonic Geological movements on a planet or moon, driven by internal forces that produce folding and faulting.

Terrestrial planet A planet whose composition is mainly rocky: Mercury, Venus, Earth, and Mars.

Ultraviolet (UV) Radiation with wavelengths just shorter than visible light.

Variable star Any star whose brightness appears to change; periods may range from minutes to years.

Volcanic Geological activity driven by the internal heat of a planet or moon.

Wavelength The distance between two successive waves of energy passing through space.

White dwarf The small, very hot remnant of a star that has evolved past the red giant stage.

X rays Radiation with wavelengths between ultraviolet and gamma rays.

Zodiac The 12 constellations straddling the ecliptic through which the Sun, Moon, and planets appear to move during the year.

Index

Numbers in *italics* indicate illustrations

CAPTIONS

Pages 2–3: The observatory domes on Mauna Kea, Hawaii.
Pages 4–5: A false-color view of the galaxy M 81.
Pages 8–9: A Hubble Space Telescope image of NGC 2440, a planetary nebula.
Pages 14–15: Part of Valles Marineris on Mars.
Pages 16–17: A combined radio and infrared image of a gravitational lens.
Pages 18–19: A medieval depiction of the zodiac and Solar System.
Pages 40–1: A space shuttle astronaut takes a walk in space.
Pages 68–9: M 51, the Whirlpool Galaxy.
Pages 70–1: Comet Hale-Bopp

photographed over Finland, 1997.
Pages 96–7: The James Clerk Maxwell Telescope, with Comet Hale-Bopp, Mauna Kea, Hawaii, 1997.
Pages 116–17: NGC 4676, two interacting galaxies in the constellation Coma Berenices.
Pages 118–19: Jupiter's Moon Io, photographed by the Galileo spacecraft.
Pages 170–1: A Hubble Space Telescope image of the galaxy NGC 4314.
Pages 204–5: A Hubble Space Telescope image of the nebula NGC 604.
Pages 220–1: A false-color image of the four stars called the Trapezium, within M 42, the

Orion Nebula.
Pages 224–5: An ultraviolet image of the Sun, taken by the SOHO satellite.
Pages 238–9: An infrared image of Jupiter, showing the impact of a fragment of Comet Shoemaker-Levy 9.
Pages 260–1: The Centaurus cluster of galaxies.
Pages 312–13: Dust and gas clouds near the star Rho Ophiuchi.
Pages 358–9: Crux, the Southern Cross.
Pages 376–7: A Hubble Space Telescope image of galaxies known as the Hickson Compact Group.
Pages 414–15: A false-color image of M 8, the Lagoon Nebula.

Contributors and Acknowledgments

CONSULTANT EDITOR

Robert Burnham is a science writer and editor specializing in astronomy and earth science. A former editor-in-chief of *Astronomy* magazine, he is the author of *The Star Book*, *The Reader's Digest Children's Atlas of the Universe*, *Great Comets*, and *Exploring the Starry Sky*. He is also the co-author of *Reader's Digest Explores Astronomy*. Robert has been an amateur astronomer since the 1950s, mainly observing the Moon and planets with his backyard telescope. He also enjoys following developments in cosmology.

CONTRIBUTORS

Alan Dyer has been an amateur astronomer and astro-photographer for 25 years, and is one of Canada's leading astronomy writers. He is co-author of *The Backyard Astronomer's Guide* and is a regular contributor to the *Observer's Handbook*, published annually by the Royal Astronomical Society of Canada. He is a former associate editor of *Astronomy* magazine and is currently a contributing editor for *Sky & Telescope* and *SkyNews*.

Jeff Kanipe is a science journalist, author, and lecturer. He is the author of *A Skywatcher's Year*

and has served as managing editor for *Astronomy* magazine and editor-in-chief of *StarDate* magazine. He conceived and wrote a daily skywatching column for SPACE.com for several years. He has also written for *New Scientist*, *Sky & Telescope*, *Astronomy*, *The World and I*, and the Earth & Sky and StarDate radio programs.

Weldon Owen would like to thank the following people for their assistance in the production of this book: John O'Byrne, Marney Richardson, Peta Gorman and Puddingburn Publishing Services (index).

Applied Physics Laboratory 147c; Lauros Giraudon 18–19; Lauros-Giraudon/Bridgeman Art Library 338tl; Kim Gordon/SPL/photolibrary.com 386c, 389t; Otto Hahn/Peter Arnold/AUSCAPE 137b; Erich Lessing/AKG London 24; Dr Jean Lorre/SPL/photolibrary.com 4–5; Mary Evans Picture Library 28t, 32t, 36t, 94br, 219b; McDonald Observatory/SPL/photolibrary.com 150t; David Malin/Anglo-Australian Observatory 197t, 199t, 403t; David Malin/Anglo-Australian Telescope 182b, 194b, 195t, 325b, 328t; David Malin/IAC 391b, 410c; David Malin/ Royal Observatory, Edinburgh 382cr, 394cr; Meade Instruments Corp. 110b; Mendillo Collection 354t; Peter Menzel/APL 48t; David Miller 99l, 99r, 101b, 112t; Morton & Milton/SPL/photolibrary.com 192b; Mt Stromlo, Siding Springs/SPL/photolibrary.com 356b; Mullard Radio Astronomy Lab/SPL/photolibrary.com 181t; Museo delle Scienze, Florence /AKG London 34b; N. Carter/North Wind Pictures 23t; NASA 53, 58b, 59b, 65t, 122cl, 123t, 124b, 126b, 128c, 128t, 133b, 133t, 141t, 145t, 150t, 159b, 163t, 164b, 166t, 176b, 179, 198t, 211, 213t, 228b, 230b; NASA/JPL Photo 61c, 61t; NASA/Oxford Scientific Films/Auscape 40–41c; NASA/SPL/ photolibrary.com 54b, 55t, 64t, 59t, 66t, 118–119c, 125b, 129, 144b, 147b, 154t, 157t, 162, 185b, 193b, 208b, 209, 216t, 217b, 217t, 224–225; Jose A. Navarro/photolibrary.com 77b; Newell Colour 151t; Jack Newton 378cl; NOAO 398bl, 403b, 406cl; NOAO/Aura/AAO 089b; North Wind Pictures 25t, 37b, 148b; Novosti/SPL/photolibrary.com 055b; David Nunuk/photolibrary.com 96–97; David Nunuk/SPL/ photolibrary.com 89t; Observatoire de Paris/Bulloz 95b; Oriental Museum, Durham University/The Bridgeman Art Library 17t, 29b, 29c, 29r; Orion Telescopes & Binoculars 10tl, 106r; Orlicka Galerie, Czech Republic/The Bridgeman Art Library 38t; Dagli Orti/The Art Archive 22b, 31b; Pacific Stock/APL 2c; David Parker/SPL/ photolibrary.com 65r, 218t; Pekka Parviainen/SPL/photolibrary.com 70–71, 98b, 169b; PhotoDisc 140t; photolibrary.com 100b; photolibrary.com/SPL 44b; Premium Stock Photography GMB/APL 74; Private Collection/The Bridgeman Art Library 31t; Ann Ronan/Image Select International 45t; Rev Ronald Royer/SPL/ photolibrary.com 246b, 249b, 321c, 388b, 395t; Royal Astronomical Society 95t, 322tl, 334t, 346tl; Royal Astronomical Society/The Art Archive 161c; Royal Geographical Society, London/The Bridgeman Art Library 29t; Royal Observatory Edinburgh/AAO 202b; Royal Observatory Edinburgh/SPL/photolibrary.com 157b; Royal Society, London/The Art Archive 36b; Royal Society, London/The Bridgeman Art Library 12b, 35b; John Sanford/SPL/photolibrary.com 139c, 254b, 344b; Ken Stepnell/photolibrary.com 134b; Scala 318tl, 350tl; SPL/photolibrary.com 405b; Dr Rudolph Schild/SPL/photolibrary.com 325t, 390c; Sloan Digital Sky Survey 203b; Space Telescope Science Institute/NASA/photolibrary.com 8–9, 16–17, 20–21b, 052b, 052c, 170–171c, 204–205, 376–377, 396–397c; Space Telescope Science Institute/SPL/photolibrary.com 184, 199b; SPL/NASA/photolibrary.com 63t, 66b, 152, 156l; SPL/photolibrary.com 43t, 46b, 49t, 55c, 68–69, 80t, 81t, 105t, 116–117c, 165b, 188t, 238–239, 260–261, 366br, 372br, 387, 389b, 393t, 405t, 412cr; Oliver Strewe 20t, 27b, 69t, 92t, 102b, 102t, 105b, 107t, 108b, 112b, 115b; Stockbyte 56b; STSI/NASA/SPL/ photolibrary.com 160; W.T. Sullivan/SPL/ photolibrary.com 101t; Tek Image/SPL/photolibrary.com 213b; John Thomas/photolibrary.com 256c; Tom Stack & Associates 252b; Tsado/NASA/Tom Stack/AUSCAPE 131b, 146t; Tsado/NOAO/TomStack/AUSCAPE 168t; TSADO/Tom Stack & Associates 185t; UCO/Lick Observatory 229, 231; United States Geological Survey 165t; UPPA/APL 138t; United States Geological Survey/SPL/photolibrary.com 64t, 142b, 161b; United States Geological Survey/NASA/TSADO/Tom Stack and Associates 14–15, 142t; United States Geological Survey/ SPL/photolibrary.com 67t, 143; USGS/Tom Stack/AUSCAPE 123c; Weldon Owen 111t; Art Wolfe/ photolibrary.com 113c; Kent Wood/photolibrary.com 80b; Westlight/APL 35t; ZEFA/APL 79br, 79cr, 79ftr, 79t.

ILLUSTRATION CREDITS
Tom Connell 12t, 17b, 42, 43b, 48–49, 51t; Lynnette Cook 121, 168b; Nick Farmer 75, 127t, 136b, 154b; Chris Forsey 69b, 84, 104tr, 104tl, 120, 146b, 174; Luigi Gallante 265, 267, 269, 271, 273, 275, 277, 279, 281, 283, 285, 289, 291, 293, 295, 297, 299, 301, 303, 305, 307, 309, 311; Lee Gibbons 180b, 181, 200; Grammacy Books, New Jersey 416, 419, 420, 422–423; David Hardy 1, 13b, 13c, 21t, 13, 60–61, 62t, 73b, 73t, 73c, 78, 79, 87b, 88, 145b, 125, 144, 117b, 208t, 117c, 117t, 122–123b, 125, 130b, 130t, 132b, 132t, 137t, 138b, 139t, 144, 147t, 149, 153, 155, 159, 167, 188b, 189b, 189t, 190b, 191, 197b, 212b; Robert Hynes 176; Rob Mancini 103, 365; Sandra Pond 22t; David Wood 81, 106br, 107bl, 107br, 108t, 166b, 169, 172, 173, 175b, 234, 240.

STAR MAPS
Wil Tirion